AIR, WATER AND SOIL POLLUTION SCIENCE AND TECHNOLOGY

SELECTED ISSUES IN EXHAUST EMISSIONS FROM AVIATION ENGINES

AIR, WATER AND SOIL POLLUTION SCIENCE AND TECHNOLOGY

Additional books in this series can be found on Nova's website under the Series tab.

Additional e-books in this series can be found on Nova's website under the e-book tab.

AIR, WATER AND SOIL POLLUTION SCIENCE AND TECHNOLOGY

SELECTED ISSUES IN EXHAUST EMISSIONS FROM AVIATION ENGINES

JERZY MERKISZ
JAROSLAW MARKOWSKI
AND
JACEK PIELECHA

nova publishers

New York

For permission to use material from this book please contact us:
Telephone 631-231-7269; Fax 631-231-8175
Web Site: http://www.novapublishers.com

Additional color graphics may be available in the e-book version of this book.

LIBRARY OF CONGRESS CATALOGING-IN-PUBLICATION DATA

Selected Issues in Exhaust Emissions from Aviation Engines / editors, Jerzy Merkisz, Jaroslaw Markowski and Jacek Pielecha
p. cm.
Includes index.
ISBN: 978-1-63117-923-5 (hardcover)
1. Selected Issues in Exhaust Emissions from Aviation Engines. Jerzy Merkisz, Jaroslaw Markowski and Jacek Pielecha
TP1180.S7P75 2010
668.4'233--dc22

2010047082

Published by Nova Science Publishers, Inc. † New York

CONTENTS

ABSTRACT

Works related to identification of exhaust emissions from aviation engines have continued since the second half of the last century. These works focus on high-thrust turbine engines. For this group testing and standardization procedures have been developed containing the admissible limits of exhaust components. Since 2007 works have been underway related to the identification of harmful exhaust components from engines of low power output that have not yet been included in the emissions legislation. These actions are particularly related to the measurements of the exhaust emissions from piston aviation engines and they are focused on the fuel applied for these engines. The book presents the results of own research work related to the issues of exhaust emissions from powertrains of aircraft and helicopters fitted with piston or turbine engines not yet included in the emission legislation. Research has been presented for turbocharged piston and jet engines aircraft. Test procedures have been presented related to the measurement of the exhaust emission under actual conditions of operation. The study presents analyses of the operating conditions of aviation engines, for which data from the on-board recording devices (flight parameters) have been used. Tests have been developed related to the engines operating under actual operating (in-flight) conditions. The methodology of the developed test has been validated based on a test dedicated for an aircraft fitted with a jet engine. The test results have been subject to a comparison with the results of tests applicable in the homologation procedures. Eventually, the Authors proposed exhaust emissions tests dedicated to individual aircraft groups.

KEY ACRONYMS

c	concentration
c_{CO}	concentration of carbon monoxide
c_{CO2}	concentration of carbon dioxide
c_{HC}	concentration of hydrocarbons
c_{NOx}	concentration of nitrogen oxides
CAFE	Cleaner Air for Europe
CARB	California Air Resources Board
CO	carbon monoxide
CO_2	carbon dioxide
DI	direct injection
e	specific emissions
e_{CO}	specific emissions of carbon monoxide
e_{CO2}	specific emissions of carbon dioxide
e_{HC}	specific emissions of hydrocarbons
e_{NOx}	specific emissions of nitrogen oxides
E	emission intensity
E_{CO}	emission intensity of carbon monoxide
E_{CO2}	emission intensity of carbon dioxide
E_{HC}	emission intensity of hydrocarbons
E_{NOx}	emission intensity of nitrogen oxides
EPA	Environmental Protection Agency
FAR	Federal Aviation Regulations
FC_C	fuel consumption coefficient
FID	flame ionization detector
G_a	air consumption
g_e	specific fuel consumption

G_e	fuel flow
HC	hydrocarbons
h_{fly}	flight altitude
ICAO	International Civil Aviation Organisation
ISA	International Standard Atmosphere
K	thrust force
k_T	model correction coefficient
LTO	landing and take-off cycle
m	mass
m_{CO}	mass of carbon monoxide
m_{CO2}	mass of carbon dioxide
m_{HC}	mass of hydrocarbons
m_{NOx}	mass of nitrogen oxides
M_o	torque
n	engine speed
N_e	effective engine power
NDIR	non-dispersive infrared
NDUV	non-dispersive ultraviolet
NO_X	nitrogen oxides
p	pressure
p_{exh}	exhaust pressure
t	time
t_{asc}	take-off and ascent times
t_{fly}	flight duration
t_{land}	descent and landing time
t_{op}	duration of ground operations
t_{steady}	steady flight time
T	temperature
u	share coefficient
v	velocity
v_{asc}	ascending velocity
v_{dsc}	descending velocity
V	capacity
w_λ	emission change coefficient
λ	air-fuel equivalence ratio

PREFACE

Research into alternative methods for analyzing emissions of hazardous compounds in exhaust gas from aircraft engines began in the 1970s. The first efforts aimed at assessing aircraft's effect on the natural environment were undertaken in the USA by the EPA in 1970. The environmental effect of toxic compounds of exhaust gas emitted by aircraft was analyzed. The said efforts contributed to publication of first standards on the emission of carbon monoxide, hydrocarbons and nitrogen oxides by the US Environmental Protection Agency in 1973. Smoke opacity standards were subsequently published in 1976.

The International Civil Aviation Organisation (ICAO) first presented its view on the negative impact of aircraft on the natural environment at the "Human Environment" conference organized by the United Nations in Stockholm in 1972. The ICAO's position was expressed in the Assembly Resolution A18-11, stating inter alia: "...in fulfilling this role ICAO is conscious of the adverse environmental impact that may be related to aircraft activity and its responsibility and that of its member States to achieve maximum compatibility between the safe and orderly development of civil aviation and the quality of the human environment ...". Subsequently, the ICAO adopted its environmental action program. Part of the program included establishment of a "study group" supporting the organization in its many tasks related to the emission of pollutants from aircraft engines. As a result of the group's efforts, a document entitled "Control of Aircraft Engine Emissions" was published in 1977. It contained guidance material in the form of certification procedures to monitor vented fuel, emission of gaseous compounds and smoke opacity for new turbojet and turbofan engines used for propulsion at subsonic speeds.

Measures undertaken by the US Environmental Protection Agency and the International Civil Aviation Organisation resulted in the emergence of two normative documents. Subsequently, efforts aimed at their alignment were undertaken. Both documents were eventually unified and published in 1981 as a separate annex to the Chicago Convention (annex 16: "Environmental Protection". It consists of the following volumes: volume I – "Aircraft Noise" and volume II – "Aircraft Engine Emissions"). The document contains standards applicable inter alia to stationary tests, measurement methods, measurement equipment and specific emission calculation methodology. In addition, the document contains emission limits for each hazardous compound in exhaust gas separately for different engine types. The original emission limits were subsequently gradually tightened in order to stimulate development of more environmentally friendly aircraft engine systems.

Most static analyses investigating the environmental impact of aircraft, as well as most development efforts carried out as part of various projects concentrate on turbine and turbofan engines used in passenger and cargo aircraft.

Thus far, not many efforts have been taken to develop guidelines for standardizing piston engines used in small general aviation aircraft and turbine engines used in helicopters. The accepted procedure methodology used to determine environmental performance of aviation engines involves stationary tests and comparisons of obtained values against the LTO test, whereby the flight phase is not addressed. The said methodology is dedicated to propulsion systems of those passenger and cargo aircraft whose operating conditions are subject to little variation.

Studies and analyses carried out as part of this publication address issues related to testing the emissions of hazardous exhaust gas compounds from engines used in aircraft and to the methodology of such tests. Therefore, this publication tackles a research problem consisting in identifying emissions of hazardous gaseous compounds in exhaust gas from aviation engines used in small general aviation airplanes, helicopters and multi-role aircraft. The basic question determining the procedure taken to solve the addressed problems is formulated as follows: how can the emission of hazardous gaseous compounds contained in exhaust gas from aviation engines be correlated with operational parameters of such engines in actual conditions?

The main objectives of this publication correspond to the key stages of its development. The first stage consisted of analyses and studies aimed at determining in-service parameters of aviation engines. To do so, it was necessary to analyze parameters recorded by on-board flight recorders used in contemporary airplanes and helicopters. In the case of airplanes with no flight recorder, in-service inspections were required. The analysis of available data made it possible to develop static tests carried out in pre-flight conditions on the hardstand, whereby test parameter values are related to corresponding in-service values. The next step involved an examination of hazardous compounds in exhaust gas from aviation engines during pre-flight tests additionally extended to include operation ranges resulting from the developed test. An analysis of the examination's results made it possible to determine the emission of hazardous compounds in the developed emission test.

For aircraft propelled with piston engines, an additional attempt was taken to develop emission models separately for each hazardous compound, based on which it would be possible to assess the emission of toxic compounds in exhaust gas from aviation piston engines in actual flight conditions.

The developed models, whose input parameters consist of information obtained at earlier stages (e.g., in pre-flight tests and in a stationary test), as well as the emissions calculated on their basis were verified by means of in-service trials. The engine and the exhaust system (as well as the formation processes of hazardous gases) were treated as an entity of unknown characteristics, for which dependencies transforming inputs into outputs were sought. An approach like that ensured the independence of measured hazardous emissions parameters from the combustion process and engine design. The two latter factors can be treated jointly, assuming that the engine is the exhaust gas generator.

This publication constitutes a concise compendium of knowledge on the relationships between the emissions of hazardous gaseous compounds in exhaust gas from aviation engines in actual conditions and the working parameters of those engines in steady conditions prescribed in the proposed study test. An additional result of this monograph is the estimation of the emissions of hazardous gaseous compounds in exhaust gas from aviation piston engines in actual conditions. Remarkably, the emission models proposed by the Authors cover all operation stages and include their assessment for piston engines. The developed emission models were generalized, thus arriving at the final model allowing one to determine the mass of any gaseous pollutant emitted by small airplanes propelled with piston engines.

Chapter 1

Overview of Emission Standards

Environmental performance of aviation engines is highly important and constitutes an impulse for the development of combustion aviation engines and aircraft designs as such. Regulations on the emission of pollutants and noise are contained in a number of international standards. This chapter presents the key assumptions of those regulations. Information contained in those regulations first of all address propulsion systems of large aircraft, while little attention is paid to propulsion systems of low power and low thrust force aircraft. Development of these aspects in the context of emission assessment becomes increasingly important from the perspective of environmental protection and development of these propulsion systems.

1.1. American Emission Standards for Aircraft

Intensive air transportation development illustrated by a steady growth in the number of aircraft and air traffic operations generates an increasing amount of atmospheric emissions of hazardous compounds. The said problem was first addressed in 1970, when the environmental effect of toxic compounds of exhaust gas from aircraft was investigated. Based on the investigation's results the need to introduce limits of hazardous compounds in exhaust gas emitted by aircraft was identified. The first exhaust gas emission standards were published by the US Environmental Protection Agency (EPA) in 1973, covering carbon oxide, hydrocarbons and nitrogen oxides. Smoke opacity standards were introduced in 1976. To correlate exhaust gas emission standards with engine types and power, seven different engine groups were identified. The division was based on the engine type and power or thrust [1, 3, 6, 7]:

- class T1 – turbojet or turbofan aircraft engines of rated power of up to 36 kN thrust,
- class T2 – turbojet or turbofan aircraft engines of rated power of 36 kN thrust or greater,
- class T3 – aircraft engines of the JT3D model family,
- class T4 – aircraft engines of the JT8D model family,
- class T5 – supersonic aircraft engines,
- class P1 – piston engines,
- class P2 – turboshaft engines used for fixed wing and rotary wing propulsion.

JT3D and JT8D engines were classified separately because a large number of such engines had been in operation for many years. Aircraft engine systems with very low power (i.e., piston engines in class P1 and turbine engines in class T1 with take-off thrust up to 27 kN, as well as turboprop engines in class P2 with maximum power below 2 MW) are exempt from emission standards. The largest emission of hazardous compounds in exhaust gas is caused by take-off and landing operations, i.e., in airport areas. To compare exhaust emission values from various types of aircraft in the airport area and to verify them against applicable emission standards, a model landing and take-off (LTO) cycle was developed. It covers aircraft operations below 914 m, i.e., during landing from the moment the plane descends below 914 m until it stops. And then during take-off, which means everything from taxiing until the plane reaches the said altitude. The aggregated emission of toxic compounds emitted in exhaust gas during the LTO cycle largely depends on the type of aircraft, as well as on the type and number of its engines. The total emission of pollutants is also affected by the LTO cycle duration. Standards on toxic emissions in exhaust gas published by the US Environmental Protection Agency prescribe emission values expressed in grams of a toxic compound per 1 kN of thrust.

On the basis of a given engine's specific fuel consumption one can determine the emission of pollutants in exhaust gas expressed in grams of toxic compounds per 1 kg of fuel – such a unit much more suitably corresponds to the combustion chamber characteristics. Importantly, each change in the engine's design leading to reduced specific fuel consumption also leads to reduction in the emission of pollutants in exhaust gas [1–8].

1.2. EMISSION STANDARDS IN FEDERAL REGULATIONS

The full name of these regulations reads Fuel Venting and Exhaust Emission Requirements for Turbine Engine Powered Airplanes. The said regulations introduce a detailed division of engines into classes (Table 1.1).

Similarly to standards put forward by the ICAO, the above standards cover specific emissions of carbon oxide, nitrogen oxides and smoke opacity. They also propose a model reference cycle characterized by means of thrust ranges (Table 1.2) and duration of each stage (Table 1.3). The presented values are required for calculating the aggregated emission of pollutants during a reference cycle, i.e., e_j/F_{00} [5, 7].

Table 1.1. Division of turbine engines into classes according to FAR 34 [5, 7]

Engine class designation	Meaning
TP	turboprop engines
TF*	dual-flow turbofan engines
T3	engines of the JT3D model family (8450–9340 daN of thrust)
T8	engines of the JT8D model family (6200–8450 daN of thrust)
TSS	includes engines designed for supersonic cruising velocities

* Engines from classes T3, T8 and TSS are included in class TF.

Table 1.2. Engine thrust ranges in the reference cycle [5, 7]

Range name	Class		
	TP	TF, T3, T8	TSS
	% of take-off thrust		
Idling, taxiing	7	7	7
Take-off	100	100	100
Ascending	90	85	65
Descending, approaching to land	undetermined	undetermined	15
Path of approach	30	30	34

Table 1.3. Engine operation times in each reference cycle stage [5, 7]

Range name	Class		
	TP	TF, T3, T8	TSS
	duration [minutes]		
Idling, taxiing	26	26	26
Take-off	0.5	0.7	1.2
Ascending	2.5	2.2	2.0
Descending, approaching to land	undetermined	undetermined	1.2
Path of approach	4.5	4.0	2.3

1.3. Global Emission Standards

To control air pollution caused by air traffic in airport areas the ICAO has introduced standards on smoke opacity (or smoke number – SN), specific emission of hydrocarbons, carbon dioxide and nitrogen oxides, as well as respective measurement procedures. The said standards apply to civil aviation subsonic aircraft with single-flow and dual-flow engines with maximum power take-off thrust exceeding 26.7 kN at ISA (International Standard Atmosphere) static sea level conditions. Since each aircraft and each power unit has its own specific exhaust gas emission profile, the model or representative measurement cycle has been introduced. The said cycle, referred to as the LTO cycle in the standard, consists of the following stages [6, 7]:

- approach to land at an altitude of 914 m,
- landing run and taxiing on the runway,
- taxiing to the runway and ascent up to 914 m.

The value of specific emission (expressed in grams per 1 kN of thrust) is determined using the following formula [6, 7]:

$$e_j\ [\mathrm{g/kN}] = E_j\ [\mathrm{g/kg_{fuel}}] \cdot g_e\ [\mathrm{kg_{fuel}/kNh}] \cdot t\ [\mathrm{h}] \tag{1.1}$$

where:

E_j – emission factor (number),
g_e – specific fuel consumption,
t – time.

The formula (1.1) shows that the total mass (expressed in grams) of substances emitted (e_j) as pollution during an LTO cycle is compared against the take-off thrust measured at the sea level. Furthermore, the following time limitations for the cycle are provided for in the formula (1.1): 19 minutes for idling and taxiing, 0.7 minute for take-off run, 2.2 minutes for ascending to an altitude of 914 m and 4 minutes for approaching to land.

The following engine settings have been assumed for the LTO test: 7% of maximum take-off thrust for idling, 100% of maximum take-off thrust for take-off, 30% of maximum take-off thrust for path of approach and approaching to land.

Values of specific emissions for each cycle stage are aggregated to calculate total emission for a full LTO cycle in relation to the take-off thrust, marked as e_j/F_{00} [g/kN].

ICAO standards also apply to smoke opacity expressed by means of the opacity number D, determined using the following formula:

$$D = 83.6\,F_{00}^{-0,274} \tag{1.2}$$

The intention behind this limitation is to eliminate visible opacity of gases exhausted from the engine. Reduction in smoke opacity in line with the growth of thrust is explained by the fact of reducing the concentration of particulate matter in a greater stream of exhaust gases.

Concluding on issues related to emission standards applicable to exhaust gas from aircraft engines one should stress that these standards apply to turbine engines. Furthermore, procedures for identifying and assessing emission parameters involve examinations carried out as part of stationary tests designed to evaluate emissions in the vicinity of airports. Therefore, they fail to comprehensively address emissions generated during the flight [6, 7].

REFERENCES

[1] Assessment and Standards Division Office of Transportation and Air Quality U.S. Environmental Protection Agency and AEE-300 - Emissions Division Office of Environment and Energy Federal Aviation Administration, Recommended Best Practice for Quantifying Speciated Organic Gas Emissions from Aircraft Equipped with Turbofan, Turbojet, and Turboprop Engines, EPA-420-R-09-901, 2009.

[2] Assessment and Standards Division Office of Transportation and Air Quality U.S. Environmental Protection Agency. Control of Air Pollution from Aircraft and Aircraft Engines; *Final Emission Standards and Test Procedures, Summary and Analysis of Comments*. EPA-420-R-12-011, 2012.

[3] Environmental Protection Agency, Control of Air Pollution from Aircraft and Aircraft Engines; Proposed Emission Standards and Test Procedures; Proposed Rule. Part I. *Federal Register*, Vol. *76*, No. 144, 2011.

[4] Environmental Protection Agency, Control of Air Pollution from Aircraft and Aircraft Engines; Proposed Emission Standards and Test Procedures; Proposed Rule. Part II. *Federal Register*, Vol. *76*, No. 144, 2012.

[5] Federal Aviation Administration, *Federal Aviation Regulations*, Part 34.

[6] International Civil Aviation Organization. (2008). Annex 16: *Environmental Protection*, Vol. *II*, Aircraft Engine Emissions.

[7] Kotlarz, W. (2004). *Turbine driving systems as pollution sources at military airports*. Dęblin: Air Forces Academy.

[8] World Health Organization. (WHO). *Health effects of transport-related air pollution*. Geneva 2005.

Chapter 2

INVESTIGATIONS INTO THE EXHAUST EMISSIONS FROM AVIATION ENGINES

The Authors' attempt at assessing toxic emissions in exhaust gas from aviation piston engines fitted in small aircraft is carried out in a number of steps. In the first stage, an analysis of operation conditions of selected aircraft was carried out, based on which an emission test procedure was developed. Such a test can be carried out on the hardstand, because its parameters are similar to operation parameters taking place during actual operation of small airplanes. Subsequently, the test was used to determine emission of pollutants. In the next step, models were developed for the purpose of determining the mass of pollutants on the basis of the stationary test results, adjusted for flight conditions and aircraft characteristics. The results calculated on the basis of the said models were verified by means of emission measurements in actual flight conditions.

2.1. INTRODUCTION

Means of transportation are a fine example of contemporary goods that are indispensable on the one hand, but cause environmental impacts that need to be reduced on the other. Advanced engineering techniques and material technologies make it possible to deliver vehicles whose manufacturing has a low environmental footprint. However, their impact on the environment and on humans caused by their operation is of equal importance. In most cases, mechanical energy required for propulsion of all vehicles in land, sea and air transportation is derived from combustion engines. Given the function of their intended use and the specific nature of energy conversion processes, such engines require particular attention. Both in terms of regular checks of the technical condition of all systems affecting the quality of thermal processes taking place within the engine, as well as with regard to the exhaust gas discharge system.

At the current state of manufacturing and material technologies and techniques used in combustion engines, the engine's efficiency with respect to converting energy from fuel to mechanical energy used for propulsion continues to be insufficient. The engine's overall efficiency is directly correlated with fuel consumption. Efforts aimed at enhancing the overall efficiency require improvement of the engine's in-service performance, which weakens its environmental performance by increasing the amount of hazardous substances in exhaust gases. As a matter of consequence, vehicle makers are inspired to look for new ways for exhaust gas aftertreatment. In the automotive industry the above trends have given rise to the idea of downsizing and introducing exhaust gas treatment systems.

In the 1970s, three way catalytic converters were first introduced to spark ignition engines. The operating principle of those converters consists in efficient oxidation of carbon monoxide and hydrocarbons and in reduction of nitrogen oxides, thus ensuring compliance with emission standards. In fact, the said principle has not changed since the converters' introduction. In compression ignition engines, ensuring compliance with today's emission standards requires the use of a treatment system consisting of a hydrocarbon and carbon monoxide oxidation catalyst, selective catalytic reduction system (to reduce nitrogen oxides) and a particulate filter. Apart from exhaust gas aftertreatment systems, contemporary combustion engines also feature multiple systems controlling the combustion process. Their operating principle involves modification of charge conditioning processes. The most important of such systems include exhaust gas recirculation and electronic fuel supply, making it possible to split the dose of fuel delivered to the cylinder in each engine cycle. One should also mention valvetrain systems that make it possible to control the whirling motion of the agent in the cylinder, ensuring proper stratification of the charge, thus optimizing its combustion.

Problems linked to the pursuit of high overall efficiency and low fuel consumption also apply to propulsion systems used in aircraft. Particular attention is paid to turbine engines used in turboprop, turbofan and jet airplanes and in helicopters. Most of them are subject to emission standards of hazardous compounds from exhaust gas provided for in Annex 16 ("Environmental Protection") of the Chicago Convention of the International Civil Aviation Organisation. Pollution emission standards do not apply to piston engines and turbine engines up to 2 MW and thrust force up to 26.7 kN. As a result of the said exemption, the development of aircraft piston engines in terms of reducing their negative environmental impacts has been stalled. Exclusion of this type of propulsion systems from environmental standards has resulted in lower investments by aircraft engine manufacturers in developing such systems. An additional cause of stalled development is the necessary aircraft engine certification, which constitutes an important cost for the manufacturer. If the number of manufactured units is low, certification costs become significant. The said number strongly depends on the number of new aircraft delivered to general aviation customers. Worldwide, general aviation is defined as all air traffic excluding scheduled and state flights. Thus, the term general aviation refers to all types of aircraft, ranging from balloons to jet planes used for private or business purposes. The largest group of general aviation aircraft are small and medium airplanes powered with piston and turbine engines. The supply of those planes is closely related to market demand (Figure 2.1).

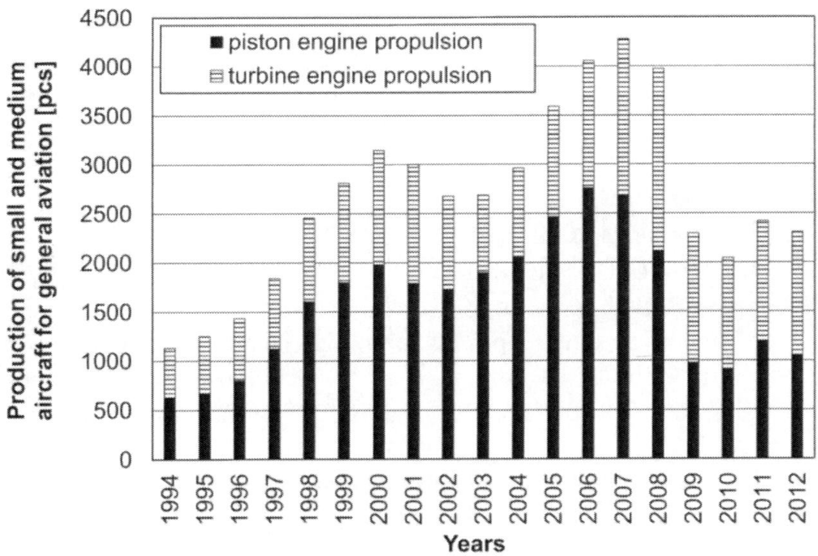

Figure 2.1. Small and medium aircraft for worldwide general aviation manufactured from 1994 to 2012 [19].

In 1994, approximately 1130 small and medium general aviation planes were manufactured, 55% of which were powered with piston engines. The period 1994 to 2000 saw a steadily growing demand averaging 20% p.a. In 2000, approximately 3150 units were manufactured, and the share of airplanes powered with piston engines grew to approximately 60%. From 2000 to 2003 the production declined slightly, as a result of which the output in 2003 was approximately 2690, with the share of piston engine powered aircraft amounting to 70%. After 2003 the production volume picked up again and kept growing until 2007, when approximately 4280 units were built, 63% of which were powered with piston engines. In the period from 2008 to 2010 general aviation aircraft production shrank dramatically, mostly as a result of the world economic crisis which – given its specific economic set-up – significantly affects general aviation. In 2010 approximately 2050 airplanes were built, and the share of those powered with piston engines went down to 45%. A gradual growth in the demand for small aircraft has been observed since 2011, but the share of piston engine planes has remained at the level of approximately 47%.

Today, the prospects for the general aviation aircraft industry suggest that over the next three years the production will reach the 2006 level, and the share of airplanes with piston engines will be approximately 45%. It is expected that the growth track will continue at least until 2020 and that the annual production volume will be approximately 6000 units, 35% of which will use piston engines. The above statistics are based on an assumption that today's technical level of piston engines will be maintained. Given the aggregated number of piston engine aircraft manufactured from 1994 to 2012 (approx. 30 000 units), this may be a serious threat for the natural environment. The above thesis is corroborated by aviation fuel consumption statistics from the years 2000 to 2012 with a 2032 forecast (Figure 1.2). The annual consumption of Avgas dedicated to piston engines is approximately 700 thousand tons, with a discernible downward trend in annual consumption towards 600 thousand tons, expected to continue until 2032. On the other hand, the consumption of Jet Fuel dedicated to turbine engines clearly grows and is anticipated to constitute approximately 95% of all fuel used in general aviation in 2032.

The above forecasts may prove true unless environmental standards on exhaust gas emissions are introduced and stimulate the development of piston engines for the aviation industry. The need for changes in propulsion systems used in small general aviation aircraft is very urgent. It is related to many areas, such as replacement of Avgas 100 and Avgas 100LL with another fuel free of lead, bromine and sulfur compounds. Such a fuel would make it possible to use three way catalytic converters, thus allowing significantly reduced emissions of carbon oxides, hydrocarbons and nitrogen oxides. Removal of lead tetraethyl from fuel requires structural and material changes in the design of piston engines used in airplanes. The use of a catalytic exhaust gas aftertreatment system requires the excess air coefficient λ to have a value close to the combustion of stoichiometric mixture $\lambda \approx 1$, which involves the use of electronically controlled fuel supply systems. The use of such systems may contribute to further optimization of the combustion process and the propulsion system as such.

Striving towards a new composition of fuel for aircraft with piston engines is reasonable not only because it stimulates engine development, but also with a view to reduced environmental impacts. A 2007 analysis carried out by the US Environmental Protection Agency (EPA) coupled with 2012 statistics on annual Avgas consumption show that the annual amount of lead discharged to the atmosphere is approximately 300 tons (Figure 2.3). It is expected that without technological improvements atmospheric emissions of lead will remain unchanged until 2032.

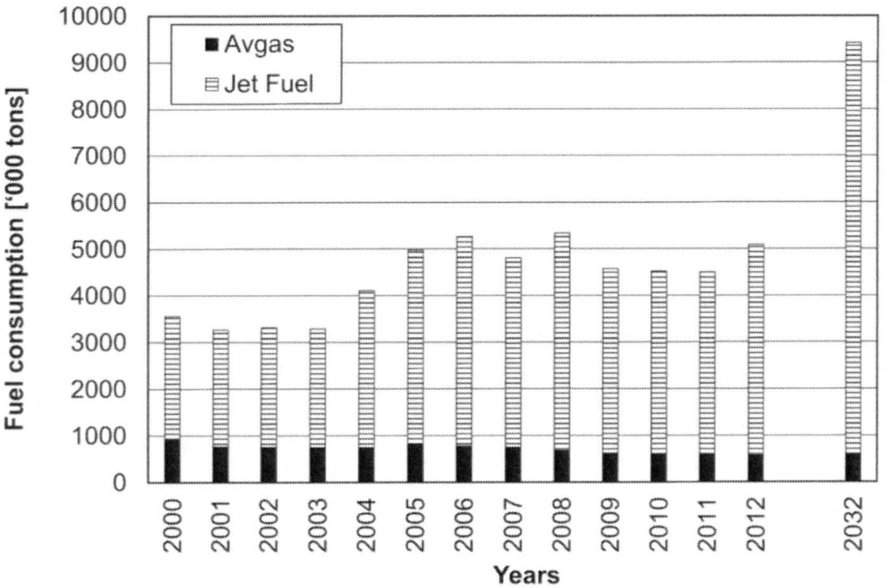

Figure 2.2. Worldwide fuel consumption by general aviation aircraft (2000-2012 with a 2032 forecast) [19].

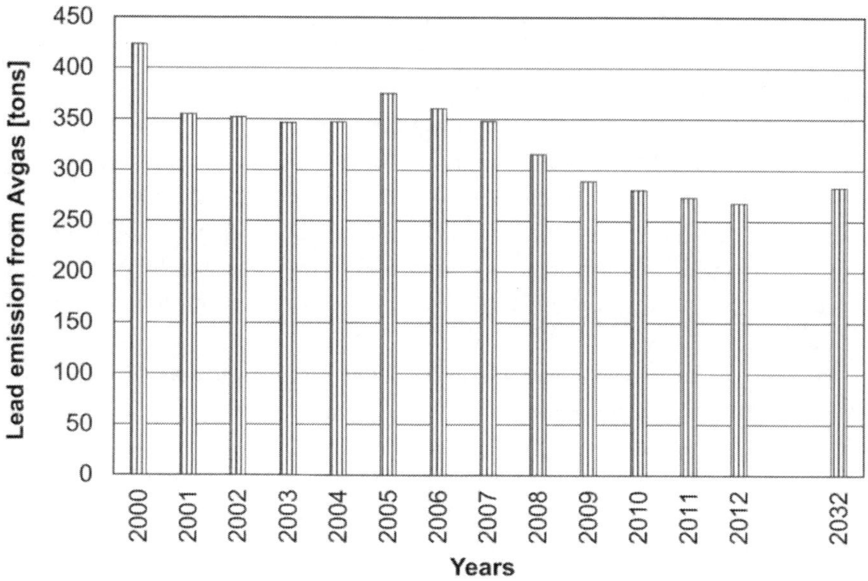

Figure 2.3. Worldwide annual emission of lead from general aviation piston engines (2000-2012 with a 2032 forecast) [19, 49].

To improve environmental performance of general aviation propulsion systems based on piston engines it is necessary to develop more innovative technologies. The starting point in this respect is an analysis of gaseous emissions of hazardous compounds contained in exhaust gas in terms of the performance of piston engines. In the case of small aircraft propelled with piston engines, no efforts at investigating in-service exhaust emissions had been taken until 2007, mostly because the available measurement equipment was too large. By now the said issue has not been fully solved and analysis of hazardous emissions from aviation piston engines in actual operating conditions is possible only for airplanes providing enough space and capable of carrying the weight of the instrumentation. Therefore, emission models must be developed in order to identify emissions of hazardous compounds from small aircraft.

Currently there exist few solutions for modeling hazardous substances emitted in exhaust gas from aircraft propelled with piston engines, and their characteristics are closely related to the landing and take-off cycle (LTO) test provided for in ICAO standards for assessing emissions from large aircraft with turbine engines. The LTO test models emissions in the airport area and excludes the flight phase [1–5, 7–17, 22, 24–27, 42–48, 50–52, 55–59, 61–64].

Available literature presents existing models for hazardous emissions in exhaust gas from other means of transportation. The presented characteristics of total emissions models are approximations over a range of considered physical values of empirical tests. Therefore, the most difficult task consists in providing reliable data for analyzing the emission model, particularly because of the high degree of generality of official transportation statistics. Therefore, emission models are too imprecise, and inaccuracies in the model's parameters are multiplied when determining the final values. Therefore, this publication aims at developing hazardous compounds' emission models dedicated to small general aviation aircraft with piston engines as well as at identifying their in-service emission of hazardous gaseous compounds contained in exhaust gases.

2.2. Methodology of Proceedings

A number of general aviation airplanes were selected for the purpose of the study, based on their availability in and representativeness for the Polish fleet of general aviation aircraft [28]. The analyzed airplanes and their key parameters are presented in Table 2.1. The selected airplanes are powered by air-cooled radial, in-line and opposed-piston engines, constituting a representative group of aviation piston engine designs. The study exclusively covers spark ignition engines. Older type engines were fitted with carburetors, while Continental engines used in Cirrus SR 20 and SR 22T were fitted with multipoint indirect injection systems. All engines were fitted with doubled ignition systems.

The PZL-104 Wilga 35 was powered with a 10.16 dm^3 9-cylinder star AI-14RA engine. The engine was fitted with a single speed mechanically driven radial compressor. The drive of the propeller W530-D11 was based on a planetary transmission of a gear ratio 0.787 : 1. The Cessna 152 II was powered with a 3.8 dm^3 4-cylinder Lycoming 0-235-L2C engine (opposed-piston design). The Zlin-142 M was powered with a 5.97 dm^3 6-cylinder Avia M 337 AK (M 337 C) engine. This in-line engine was supercharged by means of a radial compressor. The Cirrus SR20 was powered with an naturally aspirated 5.9 dm^3 6-cylinder Continental IO-360-ES engine (opposed-piston design). Other Cirrus SR22T airplanes were propelled with 9.05 dm^3 6-cylinder Continental IO-550-N piston-opposed engines. These engines were supercharged with a turbocharger and fitted with intercoolers. Engines used to propel Cirrus SR22T planes are of the same type but have different designs of their turbochargers.

Table 2.1. Aircraft covered by the study

Airplane	Key parameters
PZL-104 Wilga 35 (SP-AHS)	
	Length/span/height: 8.10/11.12/2.96 m Lifting surface: 15.50 m^2 Maximum speed: 194 km/h Maneuvering speed: 157 km/h Minimum speed: 60 km/h Gross weight / empty weight: 1230/850 kg Engine: AI-14RA, radial, 9-cylinder Engine power: 194 kW at 2500 rpm Engine capacity: 10.16 dm^3 Airscrew: W530-D11 – wooden, two-blade, adjustable pitch
Cessna 152 II (SP-NIC)	
	Length/span/height: 7.34/10.11/2.59 m Lifting surface: 14.80 m^2 Maximum speed: 276 km/h Maneuvering speed: 198 km/h Minimum speed: 63 km/h Gross weight / empty weight: 758/515 kg Engine: Lycoming 0-235-L2C, opposed-piston, 4-cylinder Engine power: 86 kW at 2800 rpm Engine capacity: 3.8 dm^3 Airscrew: McCauley Accessory Division 1A103/TCM6958 fixed pitch
Zlin-142 M (SP-FCZ)	
	Length/span/height: 9.16/7.33/2.75 m Lifting surface: 13.15 m^2 Maximum speed: 230 km/h Maneuvering speed: 215 km/h Minimum speed: 102 km/h Gross weight / empty weight: 1090/730 kg Engine: Avia M 337 AK, supercharged (M 337 C), in-line, 6-cylinder Engine power: 185 kW at 3000 rpm Engine capacity: 5.97 dm^3 Airscrew: Avia V-500A – two-blade, adjustable pitch

Airplane	**Key parameters**
Cirrus SR20	
	Length/span/height: 11.68/7.92/2.71 m Wing lifting surface: 13.71 m^2 Maximum speed: 288 km/h Maneuvering speed: 250 km/h Minimum speed: 111 km/h Gross weight / empty weight: 1386/965 kg Engine: Continental IO-360-ES, opposed-piston, 6-cylinder Engine capacity: 5.9 dm^3 Engine power: 145 kW at 2800 rpm Airscrew: Hartzell – three-blade, adjustable pitch
Cirrus SR22T (SP-AVB), (SP-AVE), (SP-IES)	
	Length/span/height: 11.68/7.92/2.72 m Wing lifting surface: 13.71 m^2 Maximum speed: 339 km/h Maneuvering speed: 300 km/h Minimum speed: 111 km/h Gross weight / empty weight: 1633/1009 kg Engine: Continental IO-550-N, turbocharged, opposed-piston, 6-cylinder Engine capacity: 9.05 dm^3 Engine power: 233 kW at 2700 rpm Airscrew: Hartzell – three-blade, adjustable pitch

The study involved both older airplanes with over ten years of service, as well as newer planes (2 to 3 years old). All examined airplanes were propelled by an airscrew, which – coupled with a combustion engine – generates load for the engine. The load depends on the airscrew's rotational speed, airplane velocity, as well as atmospheric temperature, pressure and humidity, i.e., the parameters that affect the density of the medium in which the airscrew rotates. The Cessna 152 II is fitted with a fixed-pitch airscrew, while all other selected airplanes have constant speed airscrews. The pitch of constant speed airscrew can be modified, depending on operational needs. When the plane is gaining speed, the angle of incidence is decreased, which generates smaller thrust. Pitch adjustment makes it possible to adapt the airscrew's maximum propulsion efficiency to in-flight conditions. As compared to fixed-pitch airscrews, airscrews whose pitch can be adjusted depending on the plane's velocity make it possible to achieve greater propulsion efficiency over a broad range, reduce the required length of take-off and landing runs, and improve the plane's ascending performance. By optimizing thrust, airscrew pitch adjustment positively affects the performance of the entire engine-airscrew system. Constant thrust values translate into constant engine load values, which allows the system to be optimized for top engine performance points in order to improve its overall efficiency [30–37].

The above airplanes can be grouped depending on the technical sophistication of their avionics. PZL-104 Wilga 35 has the oldest avionics, followed by Cessna 152 II and Zlin-142 M. The most advanced equipment can be found in Cirrus SR20 and SR22T, which additionally feature on-board recorder of flight parameters. To understand the operational characteristics of engine parameters for the entire group of airplanes selected for the study, actual engine parameters in flight conditions were carried out for the two airplanes at either extreme of the technical sophistication continuum, i.e., for PZL-104 Wilga 35 and for Cirrus SR22T.

In the case of PZL-104 Wilga 35 it was necessary to carry out in-flight inspection and to register the key recordable engine parameters.

For Cirrus SR22T, fitted with an on-board recorder of flight parameters, an analysis of automatically recorded parameters provided by the airplane's operator was sufficient. Out of all available records, 10 registered flights were randomly selected and subjected to statistical analysis. In this manner, mean in-flight engine parameters were arrived at.

Operational parameter values were compared against engine operation parameters from pre-flight tests. The results of those comparisons were then used to build a universal, stationary emission test presented in the following subchapter.

2.3. Investigations into the Operating Parameters of Aviation Engines

2.3.1. In-Flight Test of Piston Engine Performance in PZL-104 Wilga 35

The performance of the AI-14RA engine was tested in the PZL-104 Wilga 35 plane. The plane's characteristic parameters were presented in subchapter 2.2. Collection of diagnostic information on the engine's in-flight performance values was necessary for the purpose of the study. For the AI-14RA engine, the key performance parameter (and the only parameter diagnosable in in-flight conditions) was the engine speed. The electrical signal from the engine speed transmitter (an alternating current generator) was recorded. The signal was then converted to a linear signal and recorded by means of the LogBook 360 data acquisition system, coupled with IOtech's DBK 214 analog inputs module (Figure 2.4). Subsequently, after calibration of the measurement system and scaling of the electrical signal to the engine speed value, information on the engine speed as a function of time was obtained. To obtain information on engine load despite the lack of a directly measurable signal, it was decided to record exhaust gas temperature (by means of an additionally fitted thermocouple), ambient temperature and pressure, air humidity, as well as flight parameters by means of Garmin's 18x GPS receiver (Figure 2.5).

In-flight performance of the AI-14 RA engine fitted in the PZL-104 Wilga 35 plane was measured from the start of taxiing to the take-off site until the landing. In a standard flight, a number of phases can be identified: taxiing, take-off, ascent, steady flight, approach to land, landing and taxi to apron. Depending on the flight in question, the relative duration of each phase may vary. The scope of inspection included all the above phases. The shares of taxiing, take-off, approaching to land and landing together with taxiing to the airport are relatively insignificant in the entire flight duration. The selected flight route was between the airport of Aeroklub Poznanski in Ligowiec near Kobylnica, and the town of Wągrowiec, approximately 45 kilometers from the take-off point. Therefore, the total flight distance was approximately 90 km. The actual flight trajectory was determined based on in-flight parameters recorded by the GPS receiver (Figure 2.6).

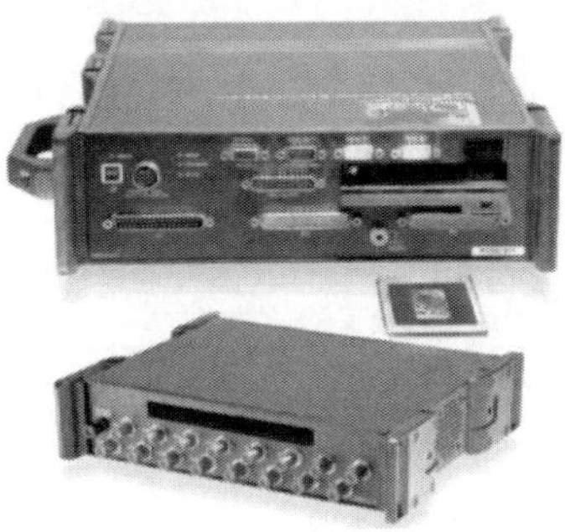

Figure 2.4. LogBook 360 data acquisition system with DBK 214 analog inputs module [32].

Figure 2.5. Garmin's 18x GPS receiver [32].

Figure 2.6. Test flight trajectory (red arrow – take-off point; white arrow – flight direction).

Out of all performance parameters recorded during the flight, the most important one was the engine speed. The following supporting parameters were measured as well: engine exhaust gas temperature and flight altitude. The selected parameters were then presented as a function of time (Figure 2.7). An analysis of the respective values of all measured parameters allows one to assess in-flight engine load changes, but does not allow one to directly calculate engine loads as such.

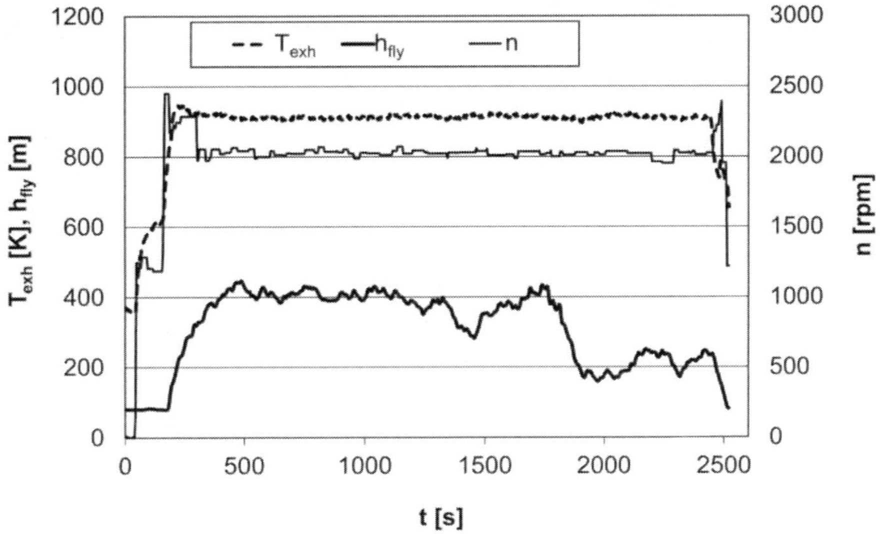

Figure 2.7. Selected engine performance parameters as a function of time recorded during the flight of PZL-104 Wilga 35.

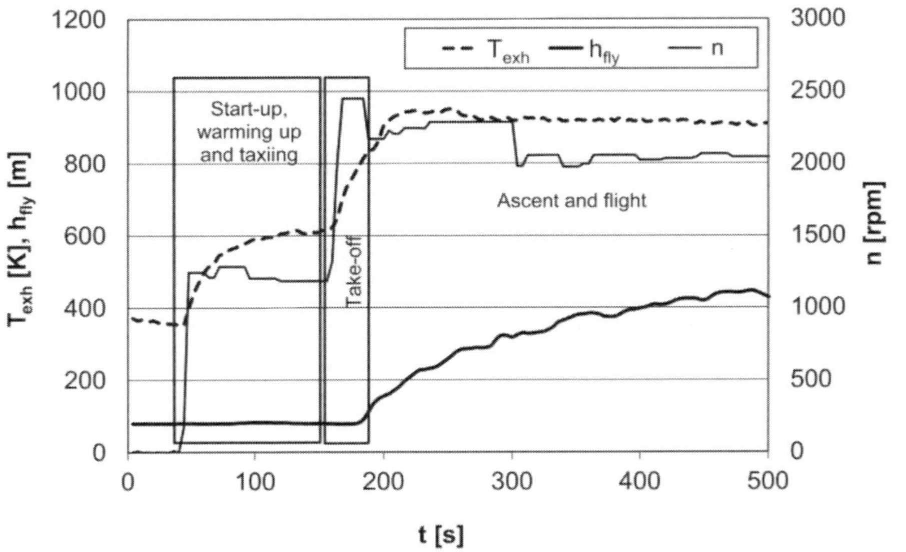

Figure 2.8. Selected in-flight engine performance parameters as a function of time – detailed view of the initial part.

Changes of engine parameters as a function of time make it possible to identify airplane operation phases. The first part of the flight record shows the engine start-up time, followed by warming-up and taxiing to the runway, and subsequently by take-off (Figure 2.8). Engine start-up and warm-up is characterized by an engine speed of approximately 1250 rpm, sustained until the beginning of take-off. The temperature of exhaust gas grows from the start-up into the warming-up phase and then stabilizes at approximately 600 K. The take-off phase is characterized by a growth in the engine speed up to its maximum value (approx. 2500 rpm). The exhaust gas temperature grows as well, which denotes a higher engine load. The end of the take-off phase is also visible (180 seconds into the flight), with a smooth transition to the first flight phase (ascent). During the ascent the engine speed is stabilized at approximately 2300 rpm, while the exhaust gas temperature grows to approximately 950 K and they stays at that level during the remaining part of the flight. The cruising engine speed is approximately 2000 rpm.

Towards the end of the record, the beginning of the approach to land procedure is noticeable (Figure 2.9). The plane begins to descend, which makes it possible to reduce the engine's load without adjusting the throttle. Reduced engine load leads to a decrease in the exhaust gas temperature. While approaching to land, the pilot reduces the plane's altitude, extends the flaps and increases the engine speed to its maximum level.

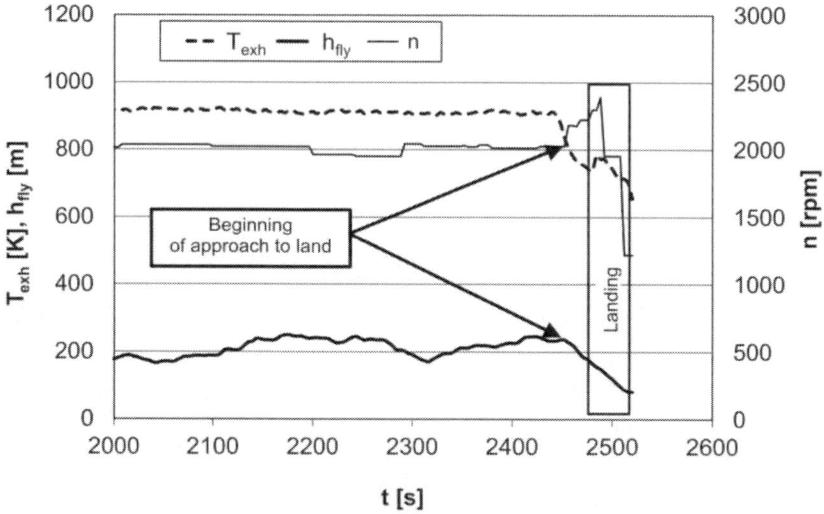

Figure 2.9. Selected in-flight engine performance parameters as a function of time – detailed view of the final part.

Table 2.2. Duration and relative share of flight operation phases

Operation phase	Time [s]	Relative share [%]
Start-up and warming up	116	4
Take-off	32	1
Ascent and flight	2292	82
Landing	28	1
Taxiing and cooling down	332	12
Total	2800	100

The above description corresponds to a standard landing procedure, aimed at ensuring full engine power in case of the need to interrupt the landing and to gain height for another approach to landing. In the case of the flight in question no such necessity arose, and therefore the engine speed was turned down to 2000 rpm as the landing procedure continued. Upon touchdown, the engine speed was further reduced to approximately 1200 rpm, being the standard value for taxiing and engine cooling until switch-off.

An analysis of the flight's duration (including taxiing after landing and cooling down the engine) shows that the entire operation took approximately 2800 seconds. The durations of all constituent phases were as follows: start-up and warming up – 116 s, take-off – 32 s, ascent and flight – 2292 s, landing – 28 s, taxiing and cooling down – approximately 332 s. The duration of each phase served as a basis for calculating its relative share in the duration of the entire flight operation, as presented in Table 2.2.

The test flight during which in-service parameters of the AI-14 RA engine were measured took 2800 seconds (or approximately 47 minutes). A flight like this can surely be considered short and it is reasonable to assume that the flight phase in most aviation operations is significantly longer.

2.3.2. In-Flight Inspection of Engine Performance in Cirrus SR22T

A randomly chosen flight from Poznań-Ławica airport to Rzeszów-Jasionka airport (some 500 km away) was selected. The flight took place in a controlled air space in accordance with the instrumental flight rules (IFRs). Flights of this kind are subject to air traffic control and are separated from one another.

From among all recorded in-flight performance parameters of the plane, the following ones were selected in order to demonstrate the engine's operating conditions: engine speed, exhaust gas temperature upstream of the turbocharger, flight altitude, airplane velocity and fuel consumption. Figure 2.10 presents the selected parameters as a function of time.

Based on the selected parameters' values, one can assess in-flight changes in engine loads, but the absolute values of such loads cannot be directly calculated. Changes in each of the parameter presented as a function of time allow one to identify the airplane's operation phases. The initial part of the record shows the engine's start-up, warming up and taxiing to the runway, followed by the take-off phase (Figure 2.11).

Engine start-up and warm-up is characterized by an unstable engine speed (ranging from 1000 to 1700 rpm), sustained until the beginning of take-off. The temperature of the exhaust gas grows from the start-up into the warming-up phase and then stabilizes at approximately 800 K. The take-off phase is characterized by a growth in the engine speed up to approximately 2500 rpm. This speed remains constant throughout the flight and is determined by the airplane's propulsion system, i.e., a triple-blade variable-pitch airscrew. The exhaust gas temperature grows in line with the engine speed, which denotes a higher engine load. Importantly, in the case of Cirrus SR22T and SR20 planes used as a means of air transportation, the settings of the engine's operational parameters remain unchanged during the ascent, until the preset altitude is reached. The take-off and ascent phase generate maximum engine loads, as demonstrated by maximum fuel consumption during these two phases. The exhaust gas temperature is 1000 K. It is in fact lower than the maximum temperature (approx. 1100 K) observed during steady flight, when engine loads vary from 70 to 80% of the maximum load. The reason behind it is the fact that during the take-off phase the engine operates on a rich mixture, with the excess air coefficient λ amounting to 0.8–0.9. Combustion of such a mixture of air and fuel is characterized by lower process temperature, which determines exhaust gas temperature. In steady flight, despite lower engine loads, the temperature of exhaust gas is higher, which is caused by a change in the excess air coefficient λ, now ranging from 0.9 to 1.0. Thus, the combustion process becomes similar to the combustion of a stoichiometric mixture, being intrinsically high-temperature.

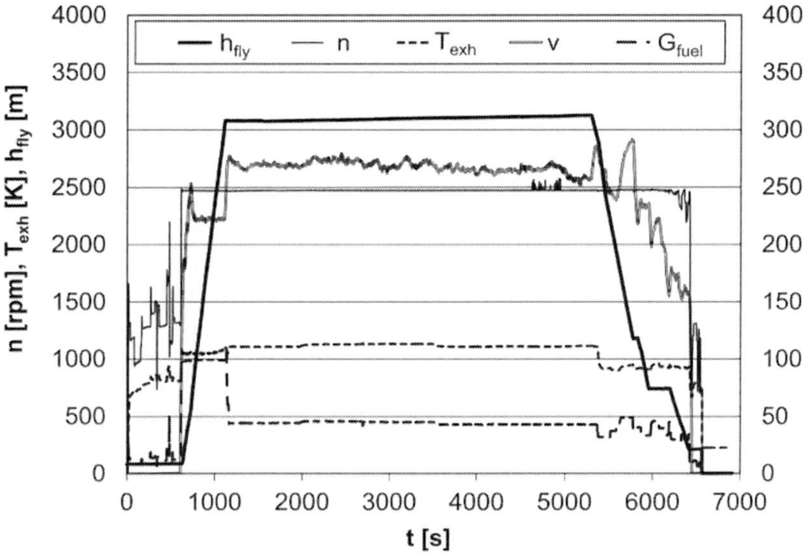

Figure 2.10. Selected in-flight parameters of the Continental IO-550-N engine as a function of time during the flight of the Cirrus SR22T (SP-IES).

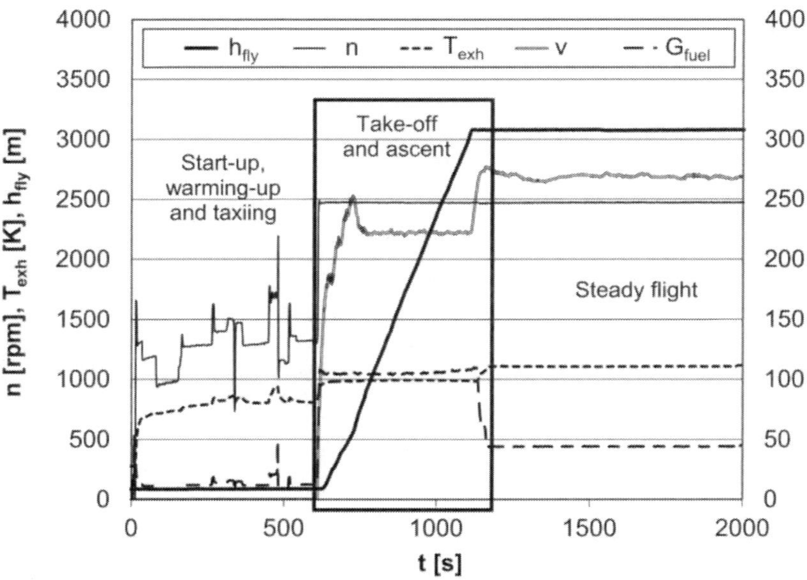

Figure 2.11. Selected in-flight parameters of the Continental IO-550-N engine as a function of time during the flight – details of the initial part.

The steady flight phase continues until the airplane begins the approach to land (Figure 2.12). As the airplane begins to descend, engine loads can be reduced. By changing the airscrew pitch it is possible to reduce the plane's altitude and velocity without changing the engine speed, which remains constant from the take-off and amounts to 2500 rpm. Lower engine load is followed by lower exhaust gas temperature, now amounting to approx. 900 K. During the landing phase, the airplane's velocity goes down, the flaps are extended and the engine speed is adjusted to maintain the required landing parameters. After landing, the plane taxis and cools down the engine that is subsequently switched off. An analysis of the time parameters of the flight in question (including taxiing after landing and cooling down the engine) shows that the entire operation took approximately 6565 seconds. The durations of all constituent phases were as follows: start-up and warming up – 600 s, take-off and ascent to 500 m – 150 s, ascent above 500 m – 385 s, steady flight – 4167 s, approaching to landing down to 740 m – 895 s, landing – 228 s, taxiing and cooling down – 140 s. The duration of each phase served as a basis for calculating its relative share in the duration of the entire flight operation. The results are presented in Table 2.3.

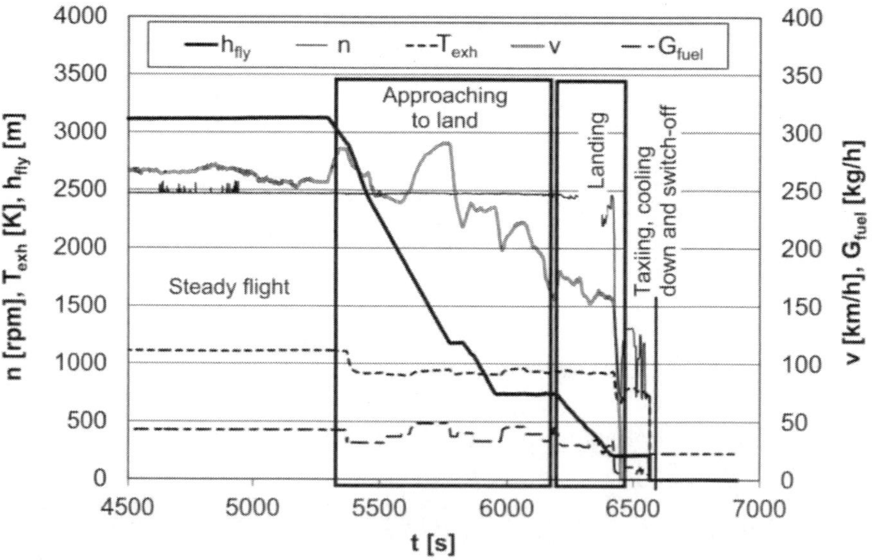

Figure 2.12. Selected in-flight parameters of the Continental IO-550-N engine as a function of time during the flight – details of the final part.

Table 2.3. Duration and relative share of flight operation phases

Operation phase	Time [s]	Relative share [%]
Start-up and warming up	600	9
Take-off	150	2
Ascent	385	6
Steady flight	4167	63
Approach to land	895	14
Landing	228	3
Taxiing and cooling down	140	2
Total	6565	100

The test flight during which in-service parameters of the Continental IO-550-N engine were measured took 6565 seconds (or approximately 110 minutes). Assuming that the ascent and approach to landing phases are constituent parts of the flight, the aggregated share of actual flight in the total operation time amounts to 83%.

The two sample flights discussed above can surely be considered short and it is reasonable to assume that the flight phase may actually be longer and take up to 3 hours, depending on the pilot's and passengers' condition and on the availability of fuel. Therefore, the relative share of the flight phase can grow up to as much as 90%.

In the case of training flights, a considerably more time is devoted to approach practice. In such flights, the relative share of the flight phase is decreased to the advantage of take-off and landing phases. From the perspective of the issue addressed by this publication it seems relevant to carry out an assessment of exhaust emissions from flights of at least 1.5 hours, i.e., general aviation business or tourist flights.

2.4. Analysis of In-Flight Engine Loads

Piston aviation engines used for propulsion purposes constitute part of the propulsion system in charge of generating mechanical energy. The energy generated in the engine is then converted to thrust, generated by the airscrew. By properly arranging these two elements to form a propulsion system based on the engine's general characteristics and the airscrew's propulsion characteristics, it is possible to ensure the airplane's propulsion exhibiting the desired performance characteristics.

The key parameters describing the engine's operating conditions include engine speed and engine load, expressed either as the torque or the mean effective pressure. The value of the engine's torque and the mean effective pressure depend on the propelling unit – the airscrew in this particular case. The airscrew rotates in a medium (atmospheric air) characterized by such parameters as: pressure, temperature, humidity affecting the air's density, and wind speed. The said parameters are reflected in loads affecting the airscrew's blades, and thus influence the load generated by the engine. The engine load is closely related to the airscrew characteristics and the ambient conditions, but its value is unimportant from the perspective of operating the plane, which is why the load is not a measured parameter. For new-generation airplanes fitted with variable-pitch airscrews, the engine load is partially used to optimize the parameters of the entire propulsion system. However, rather than measuring the engine's torque, the approximate value of the engine's power is assessed depending on the values of other parameters. To assess instantaneous engine power one can use for instance its speed, fuel consumption, air consumption, throttle position, exhaust gas temperature compared to one another by means of a suitable algorithm. For the airplanes selected for the purpose of the study presented herein it was not possible to measure the engine's torque, which is why engine power was assessed (expressed as a percentage). To that end, in newer airplanes fitted with turbocharging systems, the value of exhaust gas temperature upstream of the turbine was used. For the PZL-104 Wilga 35 airplane, exhaust gas temperature was measured at a distance of approximately 2 meters away from the exhaust valve. In this particular case it was additionally estimated that the said exhaust gas temperature in the measurement point is by approximately 150 K lower than the exhaust gas temperature in the engine's exhaust manifold, which is accounted for in the analysis. In addition, values of the engine speed and values of the excess air coefficient are determined by the fuel supply system settings. Both sets of values were compared in order to propose the equation (2.1), making it possible to assess the relative value of the engine's effective power.

$$\frac{N_e}{N_{e\max}} = \frac{T_{exh}}{T_{exh\max}} \left(\frac{1}{\lambda}\right)^{2.5} \frac{n}{38.6} \quad [\%] \tag{2.1}$$

where:

$N_e/N_{e\,max}$	relative effective power [%],
T_{exh}	exhaust gas temperature [K],
$T_{exh\,max}$	maximum exhaust gas temperature [K],
λ	excess air coefficient [–]; values for flight phases: take-off – λ = 0.8; steady flight – λ = 0.95; other operation phases $\lambda \approx 1$,
n	engine speed [rpm].

Using the proposed equation (2.1) and the measured in-flight exhaust gas temperature values in the PZL-104 Wilga 35 airplane (Figure 2.13), the relative engine power was determined (Figure 2. 14).

A similar procedure was followed to analyze in-flight performance of the Cirrus SR22T airplane. Using the proposed equation (2.1) and the measured in-flight exhaust gas temperature values (upstream of the turbine) recorded by the on-board flight recorder during the actual flight phase (Figure 2.15), the relative engine power was determined (Figure 2.16).

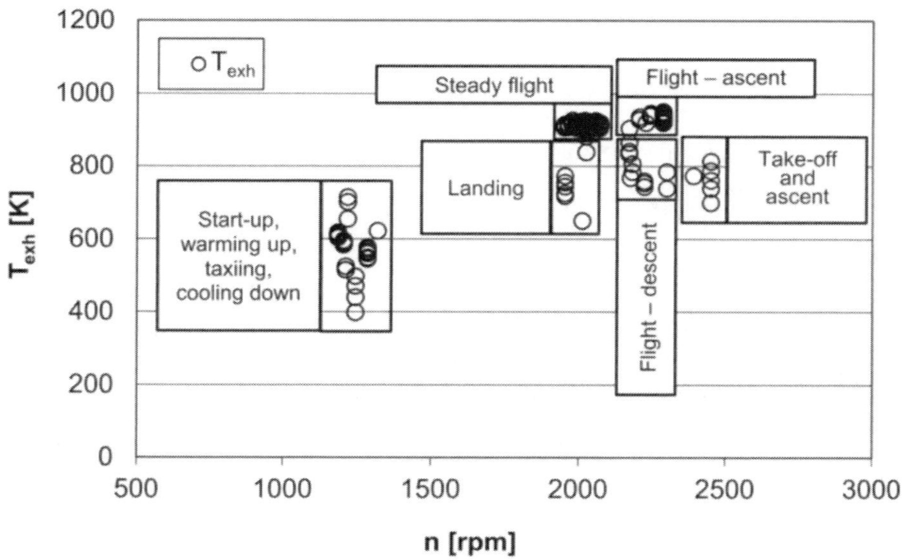

Figure 2.13. PZL-104 Wilga 35 exhaust gas temperatures and their corresponding operation areas.

An analysis of the resultant relative effective power values for different operation ranges of engines powering the PZL-104 Wilga 35 (representing older aircraft in the analyzed group of planes) and the Cirrus SR22T (representing newer aircraft) allows one to identify three main ranges of relative effective power. The first range corresponds to engine start-up and warming up, taxiing, landing and engine cooling. Relative engine power values in this range are variable, but do not exceed 60% of the maximum power. Another noticeable operation range corresponds to the steady flight phase, including ascent and descent. For this range, power values vary from 60 to 80% of the engine's maximum power. Finally, the third range corresponds to the take-off phase, whereby the engine works at 90 to 100% of its maximum power.

The operation ranges of relative power of piston aviation engines determined above are aligned with operation processes resulting from the specific nature of aviation operations and individual flight phases. On the basis of operation parameters from the flights discussed above an attempt was taken at developing a universal test for measuring exhaust emissions from piston engines used in general aviation aircraft.

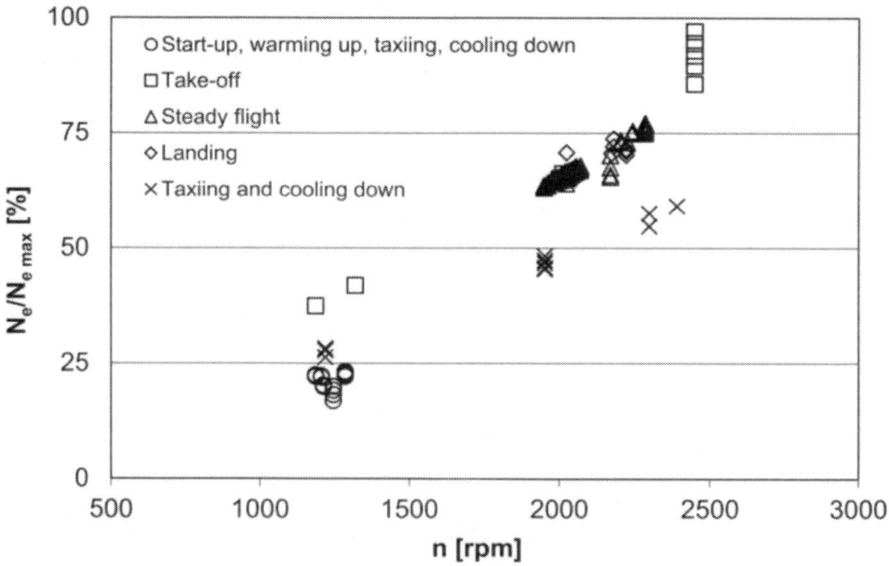

Figure 2.14. PZL-104 Wilga 35 relative effective engine power and its corresponding operation areas.

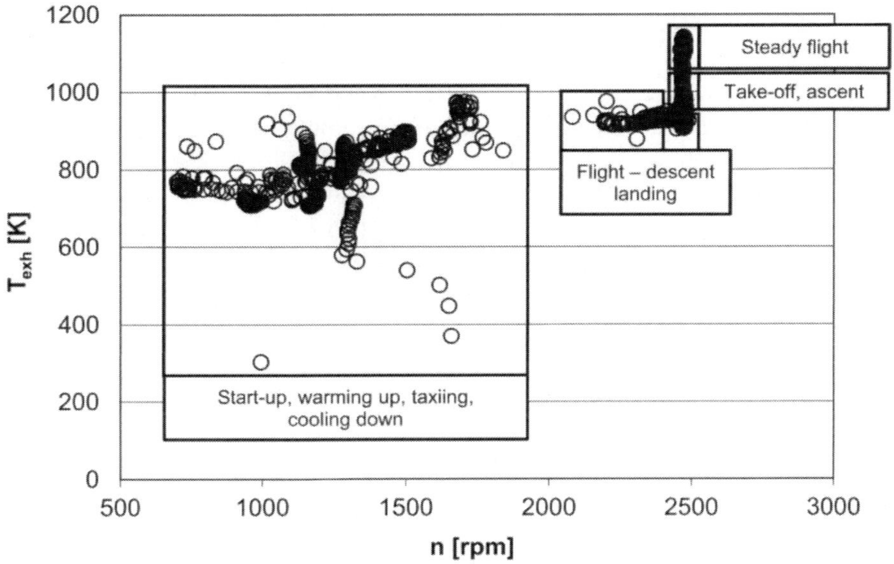

Figure 2.15. Cirrus SR22T exhaust gas temperatures and their corresponding operation areas.

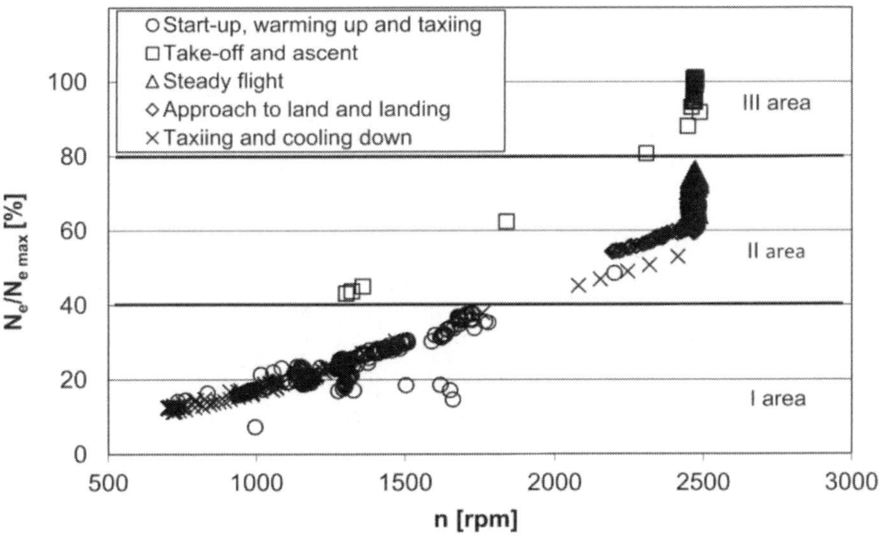

Figure 2.16. Cirrus SR22T relative effective engine power and their corresponding operation areas.

2.5. Stationary Test Design

Verification of exhaust emissions from small aircraft propulsion systems makes it necessary to measure pollution in actual in-flight conditions. For technical reasons, in many general aviation airplanes it is not possible to make such measurements. The key reason is the prohibitively low allowable load and excessively small cargo space, making it impossible to install the measuring equipment. Another difficulty is related to exhaust gas sampling. Therefore, it is necessary to develop a procedure allowing one to measure in-flight performance during a pre-flight test. The pre-flight test is characterized by the fact that it covers all settings of devices affecting engine loads, used in actual operating conditions during the flight.

In order to develop guidelines for a stationary test making it possible to determine engine emissions whose value to some extent corresponds to the actual value, actual in-flight engine loads and their respective durations were determined. On that basis it was assumed that the stationary emissions test would consist of three phases. Phase A – covering engine start-up and warming up, taxiing, landing and engine cooling. Phase B – covering engine operation parameter values recorded during the take-off. Phase C – covering steady flight including ascent and descent, except for ascent right after the take-off. Based on the determined operation parameters and measured flight durations, the relative times of individual operation phases were calculated in the previous chapter. Relative times of individual operation phases closely depend on the total flight duration; the longer the flight, the lower the share of the first and the second operation states. Flight durations in general aviation usually range between 1 and 3 hours. Therefore, one can assume that the mean operation time of small piston engine aircraft is 1.5 hours ($t_A + t_B + t_C = 5400$ s), and that assumption was borne in mind when developing the stationary emissions test carried out during a pre-flight trial.

Analyses of the engine loads distribution in in-flight conditions made it possible to identify three key operation ranges, attributed to individual test phases. To ensure the test's universality, mean relative shares of each operation phase were assumed (Table 2.4). Subsequently, the mean relative shares of operation phases were grouped together in accordance with their attribution to test phases and the following aggregated values were arrived at: phase A – 15.5%, phase B – 4.5%, phase C – 80% of the total operation time.

The resultant relative operation time shares for each operation state can be used as relative phase share coefficients in the design of the stationary test.

For each test phase, values of operation parameters should be determined. The said values will serve as test measurement points in the engine's overall characteristics. Based on operation points determined in the analysis, operation parameters were identified as relative engine speed and relative power in the test (Figure 2.17 and 2.18). For the distribution of points in the engine operation range for the PZL-104 Wilga 35 airplane, the following relative engine speed and relative power were determined: for point A – $n/n_{max} = 50\%$, $N_e/N_{e\ max} = 20\%$; for point B – $n/n_{max} = 100\%$, $N_e/N_{e\ max} = 100\%$; for point C – $n/n_{max} = 86\%$ and $N_e/N_{e\ max} = 70\%$. For the distribution of points in the engine operation range for the Cirrus SR22T airplane, the following relative rotational speed and relative power were determined: for point A – $n/n_{max} = 50\%$, $N_e/N_{e\ max} = 20\%$; for point B – $n/n_{max} = 100\%$, $N_e/N_{e\ max} = 100\%$; for point C – $n/n_{max} = 100\%$ and $N_e/N_{e\ max} = 70\%$. The predetermined operation points dedicated to each test phase for both planes are similar for points A and B, where the differences between them are not greater than 10%.

Table 2.4. List of durations and relative shares of aviation operation phases

Airplane	**PZL-104 Wilga 35**		**Cirrus SR22T**		**Mean phase share [%]**	**Test phase**
Operation phase	**time [s]**	**phase share [%]**	**time [s]**	**phase share [%]**		
Start-up and warming up	116	4	600	9	6.5	A
Take-off and ascent	32	1	535	8	4.5	B
Steady flight	2292	82	5062	78	80	C
Landing	28	1	228	3	2	A
Taxiing and cooling down	332	12	140	2	7	A
Total	2800	100	6565	100	100	–

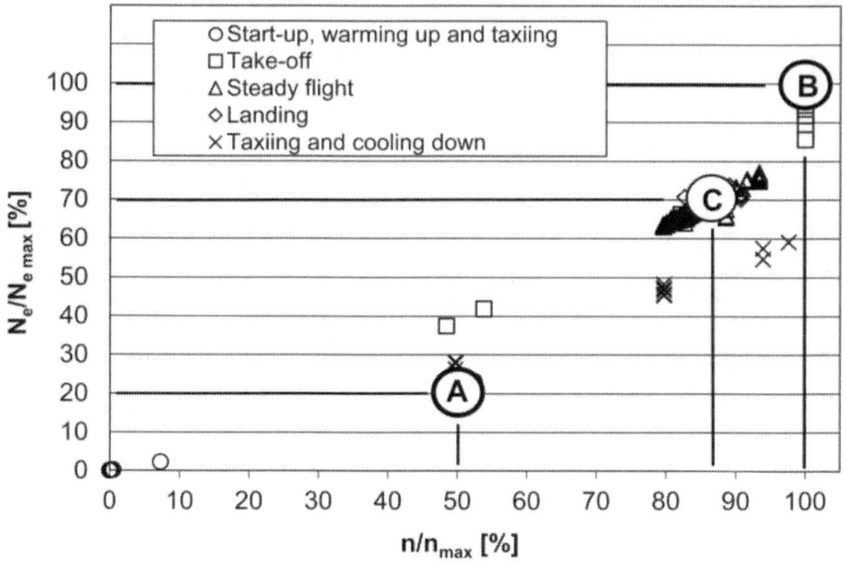

Figure 2.17. Proposed engine operation parameters for individual phases of the emission test for PZL-104 Wilga 35.

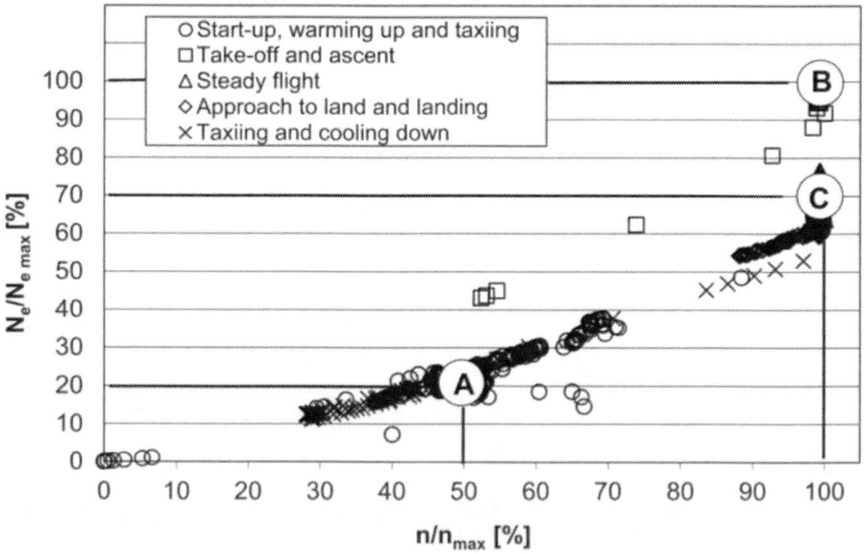

Figure 2.18. Proposed engine operation parameters for individual phases of the emission test for Cirrus SR22T.

For point C, there is a considerable difference in engine speeds of the two propulsions systems, amounting to 14%. The difference is caused by the fact that Cirrus SR22T has a variable-pitch airscrew, which is why the engine fitted on the plane is characteristic for low speed. In its turn, PZL-104 Wilga 35 has a fixed-pitch airscrew, and thus the thrust is modified by modifying its rotational speed. Therefore, the test in question can be universalized for all piston engine airscrew propulsion systems only as regards test points A and B. It is not possible for point C, due to its large share in actual airplane operation. Therefore, it is necessary to develop similar, yet separate tests dedicated to propulsion systems featuring fixed and variable pitch airscrews (Table 2.5). The difference between the proposed values of test point coordinates corresponds to the setting of the engine speed in point C and amounts to n/n_{max} = 85% for fixed-pitch airscrews and n/n_{max} = 100% for variable-pitch airscrews. Figure 2.19 shows a visualization of test point distribution, together with relative shares of each phase.

Table 2.5. Test points coordinates

Test phase	Airscrew design				Phase share coefficient
	fixed pitch		variable pitch		
	n/n_{max} [%]	$N_e/N_{e\ max}$ [%]	n/n_{max} [%]	$N_e/N_{e\ max}$ [%]	
A	50	20	50	20	0.155
B	100	100	100	100	0.045
C	85	70	100	70	0.800

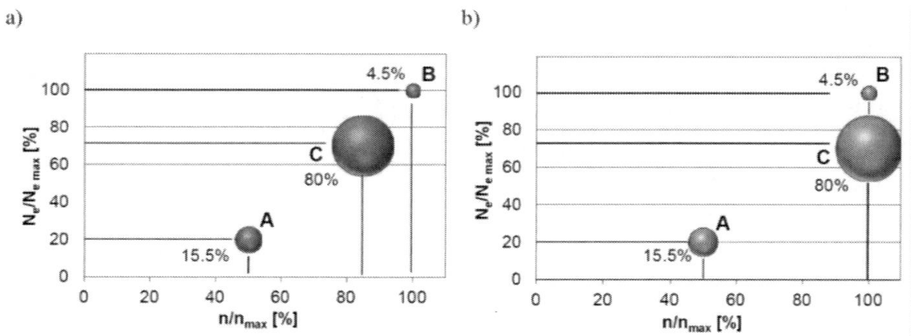

Figure 2.19. Values of engine relative speed and relative power together with phase share coefficients: a) fixed-pitch airscrew propulsion, b) variable-pitch airscrew propulsion.

The nature of the test is stationary. Therefore, the sequence and durations of all test phases is preset. The phases should be carried out in an alphabetical order (A, B, C), corresponding to the natural sequence of phases in actual aircraft operation. Therefore, the thermal state of the engine will approximately correspond to the actual state. The duration of phase A should be 4 minutes, with 2 minutes for phase B and again 4 minutes for phase C. The duration of phase B whereby maximum in-flight values corresponding to engine loads during the take-off are tested is preset at 2 minutes because in the case of a longer operation at such parameters the engine may be overheated and damaged as a result. Measurement of concentrations of individual toxic compounds in the exhaust gas should be made after all performance parameters have stabilized. Therefore, it is recommended that emission concentrations should be measured in the last minute of the test phase in question.

2.6. Tests under Actual Conditions of Operation

The content of toxic compounds in the exhaust gas of small aircraft engines was tested on the ground for the following airplanes: PZL-104 Wilga 35, Cessna 152 II, Zlin-142 M, Cirrus SR20, Cirrus SR22T (SP-AVE, SP-AVB, SP-IES). The engine's design, technological advancement and, first of all, its actual condition are of major importance when measuring exhaust emissions from piston engines. All tested airplanes were fitted with short exhaust systems, which translates into noise emitted during their operation and high exhaust gas temperatures. For structural reasons and in order to utilize the dynamic effects of exhaust gas for scavenging the cylinder, exhaust systems are divided into two parts. As a result, the tested airplanes had two exhaust pipes. The exhaust emissions were measured in one of those pipes assuming that the measured concentration of toxic compounds corresponds to the mean concentration coming from each of the cylinders. The measured exhaust gas flow rate was applied for all cylinders at the analysis stage.

The content of toxic compounds in the exhaust gas of the engines of the tested airplanes was measured on the ground during a pre-flight test. In a standard aviation operation one can identify several phases: taxiing to the runway, take-off, ascent, steady flight, approach to land, landing and taxiing to the apron. Depending on the flight in question, the relative duration of each phase may vary. The study included a pre-flight test whereby the values of operation parameters matching the proposed test phases were considered. For all operation points characterizing the phases A, B and C, the positions of engine controlling devices and airscrew pitch corresponded to their positions in in-flight conditions.

To measure the concentration of toxic compounds in exhaust gases, Sensors's Semtech DS exhaust gas analyzer was used (Table 2.6) [53]. The analyzer makes it possible to measure the concentration of carbon monoxide, nitrogen oxides, hydrocarbons and carbon dioxide. Additionally, it is fitted with a flow meter used to measure the exhaust gas flow rate (by mass). An exhaust gas intake probe is an embedded feature of the flow meter. Exhaust gas are introduced into the analyzer by means of the measurement probe and a feed-in tube sustaining the temperature of 191°C. After particulate matter is filtered, the concentration of hydrocarbons in a flame ionization detector is measured.

Subsequently exhaust gas are chilled to 4°C and the concentration of nitrogen oxides is measured – via the non-dispersive method using ultraviolet radiation making it possible to simultaneously measure nitrogen monoxide and dioxide, as well as carbon monoxide and dioxide – via the non-dispersive method using infrared radiation, and then oxygen – by means of an electrochemical analyzer (Figure 2.20).

Due to the limitations of the measurement system it is not possible to measure exhaust gas in temperatures beyond 700°C. Therefore, to carry out the test the exhaust pipes were extended to reduce the temperature to acceptable levels (Figure 2.21)

Table 2.6. Overview of Semtech DS mobile exhaust gas analyzer [53]

Measured compound	Measurement method	Accuracy
CO	non-dispersive infrared analyzer, range: 0–10%	±3%
HC	flame ionization detector, range: 0–10 000 ppm	±2.5%
$NO_x = (NO + NO_2)$	non-dispersive ultraviolet analyzer, range: 0–3000 ppm	±3%
CO	non-dispersive infrared analyzer, range: 0–20%	±3%
O_2	electrochemical analyzer, range: 0–20%	±1%
Exhaust flow meter	mass flow rate, exhaust gas temperature < 700°C	±2.5%

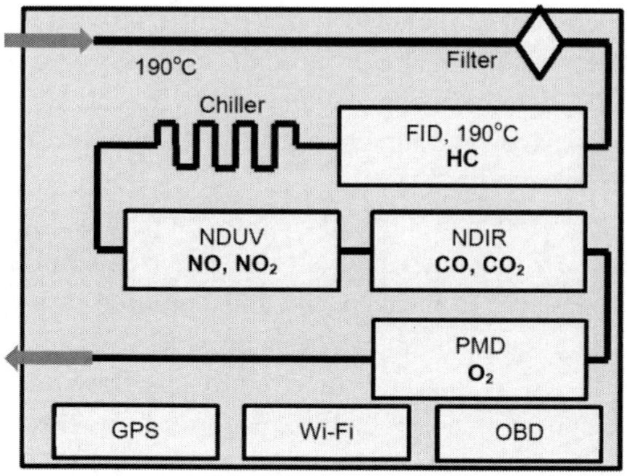

Figure 2.20. Flowchart of the Semtech DS portable analyzer showing the exhaust gas flow direction and the sequence of measurements [53, 54].

For PZL-104 Wilga 35 the exhaust system was extended by 3 meters, and for all other airplanes by 4 meters. As a result, the measurement of toxic compounds concentration took place in a location allowing one to install and secure the measurement probe and the flow meter. Measurement of toxic compounds contained in the exhaust gas from aviation piston engines carried out in stationary conditions typical of pre-flight tests included measurement of carbon monoxide, hydrocarbons, nitrogen oxides and carbon dioxide. The exhaust gas mass flow rate was measured as well. The following engine parameters were analyzed: engine speed and exhaust gas temperature. All parameters were measured as a function of time, at a frequency of 1 Hz. As a result, concentrations of each compound as a function of time were obtained. Based on instantaneous concentrations of each pollutant and the mass flow rate, the instantaneous emission rate for each compound was determined as well. All functions were subsequently divided into sections corresponding to operation phases. Mean values of recorded concentrations were calculated for each phase. Instantaneous emission rates for each phase were aggregated to obtain the total mass of a compound emitted during that phase. Subsequently, bearing in mind the duration of each phase, the average hourly emission of toxic compounds for individual phases was determined. The division of recorded changes in the measured values into operation phases was carried out bearing in mind changes in the operation parameters.

PZL-104 Wilga 35

Cessna 152 II

Zlin-142 M

Cirrus SR20

Cirrus SR22T (SP-AVB)

Cirrus SR22T (SP-AVE)

Cirrus SR22T (SP-IES)

Figure 2.21. Location of the exhaust gas intake probe during the tests.

Measurement results were presented in a graphic form and recorded in tables, separately for each tested plane:

- PZL-104 Wilga 35 – changes in toxic compounds concentration and flow rate during a pre-flight test (Figure 2.22), measurement results of toxic compounds emission in a pre-flight test (Table 2.7), results of mean hourly emissions of measured toxic compounds in a pre-flight test (Figure 2.23);

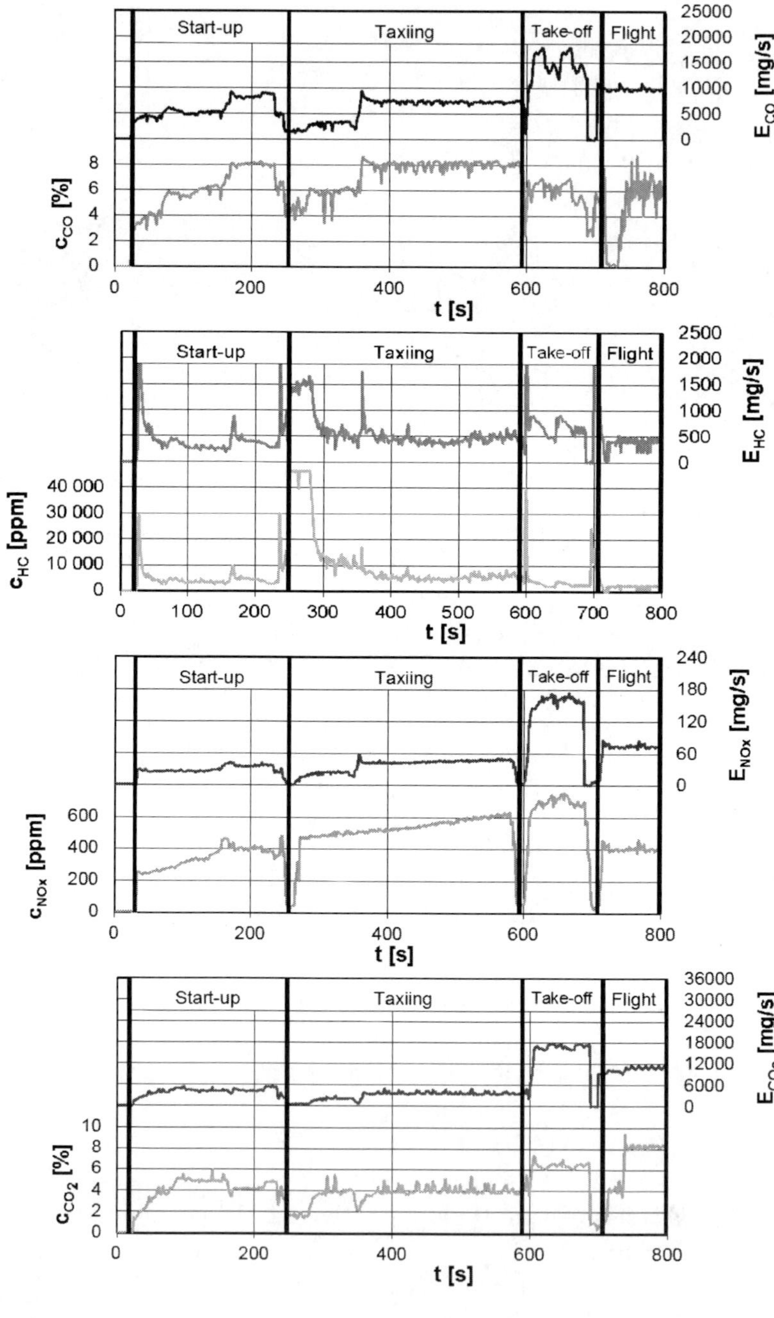

Figure 2.22. Changes in toxic compounds concentration and flow rate during a pre-flight test for PZL-104 Wilga 35.

Table 2.7. Measurement results of toxic compounds emission in a pre-flight test for PZL-104 Wilga 35

Parameter	Operation phase			
	start-up	taxiing	take-off	steady flight
Engine speed [rpm]	–	1200	2500	2000
Duration [s]	236	330	86	71
Mean concentration:				
c_{CO} [%]	5.96	7.26	6.04	5.78
c_{HC} [ppm]	6765	11 251	2904	1298
c_{NOx} [ppm]	3.04	2.82	9.57	1.04
c_{CO2} [%]	4.52	4.35	7.39	9.11
Mass [g]				
m_{CO}	1341	1969	1267	602
m_{HC}	114	188	59.0	28.8
m_{NOx}	6.40	12.2	13.0	5.25
m_{CO2}	1808	2034	2838	1558
Hourly emission [g/h]				
E_{CO}	20 458	21 477	53 054	30 515
E_{HC}	1743	2057	2470	1463
E_{NOx}	97.7	133	544	266
E_{CO2}	27 590	22 198	118 802	78 986

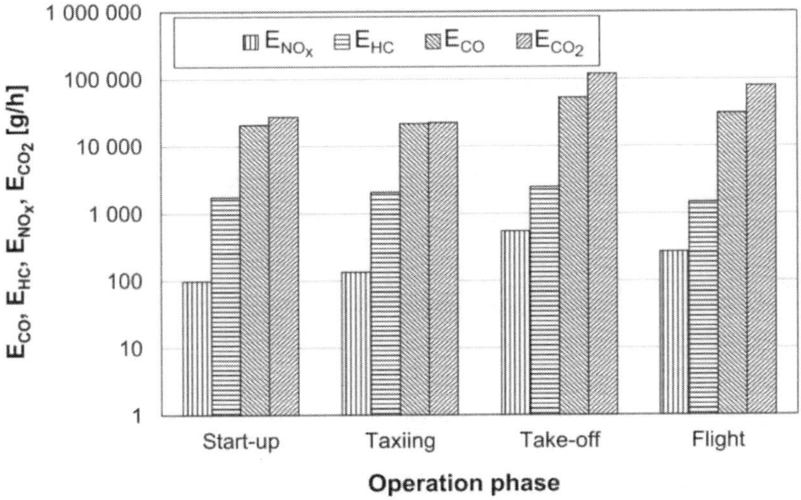

Figure 2.23. Mean hourly emissions of toxic compounds measured in a pre-flight test for PZL-104 Wilga 35.

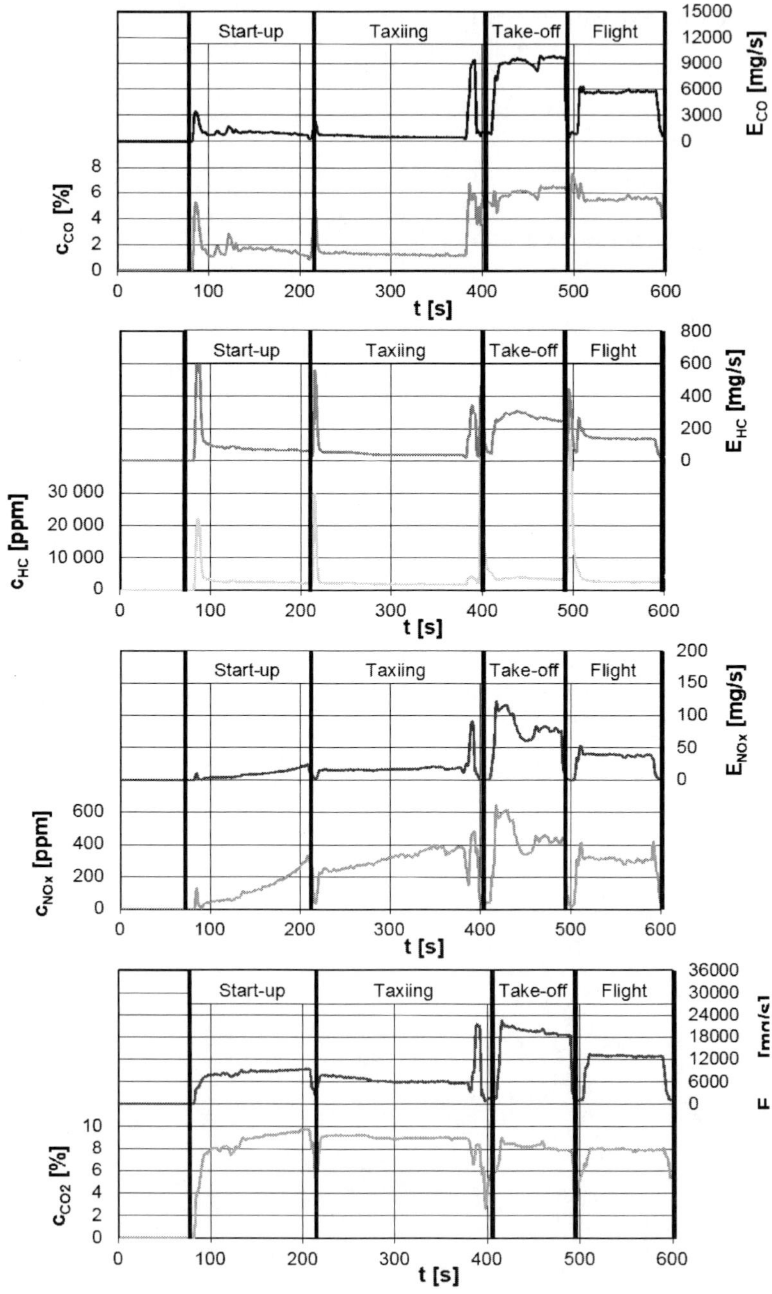

Figure 2.24. Changes in toxic compounds concentration and flow rate during a pre-flight test for Cessna 152 II.

Table 2.8. Measurement results of toxic compounds emission in a pre-flight test for Cessna 152 II

Parameter	Operation phase			
	start-up	taxiing	take-off	steady flight
Engine speed [rpm]	–	1200	2400	2000
Duration [s]	134	184	83	92
Mean concentration:				
c_{CO} [%]	1.91	1.89	7.10	6.40
c_{HC} [ppm]	3334	2679	3676	2901
c_{NOx} [ppm]	96.2	228	324	228
c_{CO2} [%]	9.24	9.74	9.51	9.07
Mass [g]				
m_{CO}	141	184	731	504
m_{HC}	13.5	12.0	21.8	13.0
m_{NOx}	1.25	3.37	6.52	3.41
m_{CO2}	1070	1241	1548	1123
Hourly emission [g/h]				
E_{CO}	3791	3594	31 721	19 731
E_{HC}	363	235	946	507
E_{NOx}	33.6	66.0	283	133
E_{CO2}	28 741	24 272	67 141	43 943

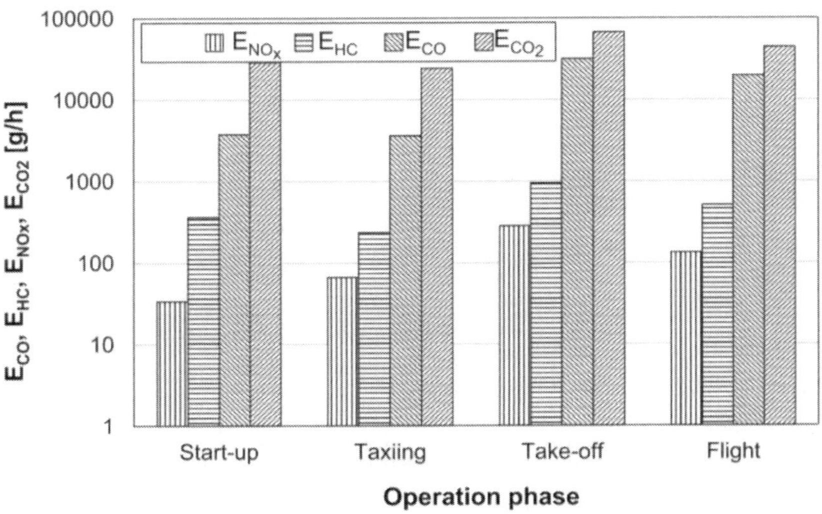

Figure 2.25. Mean hourly emissions of toxic compounds measured in a pre-flight test for Cessna 152 II.

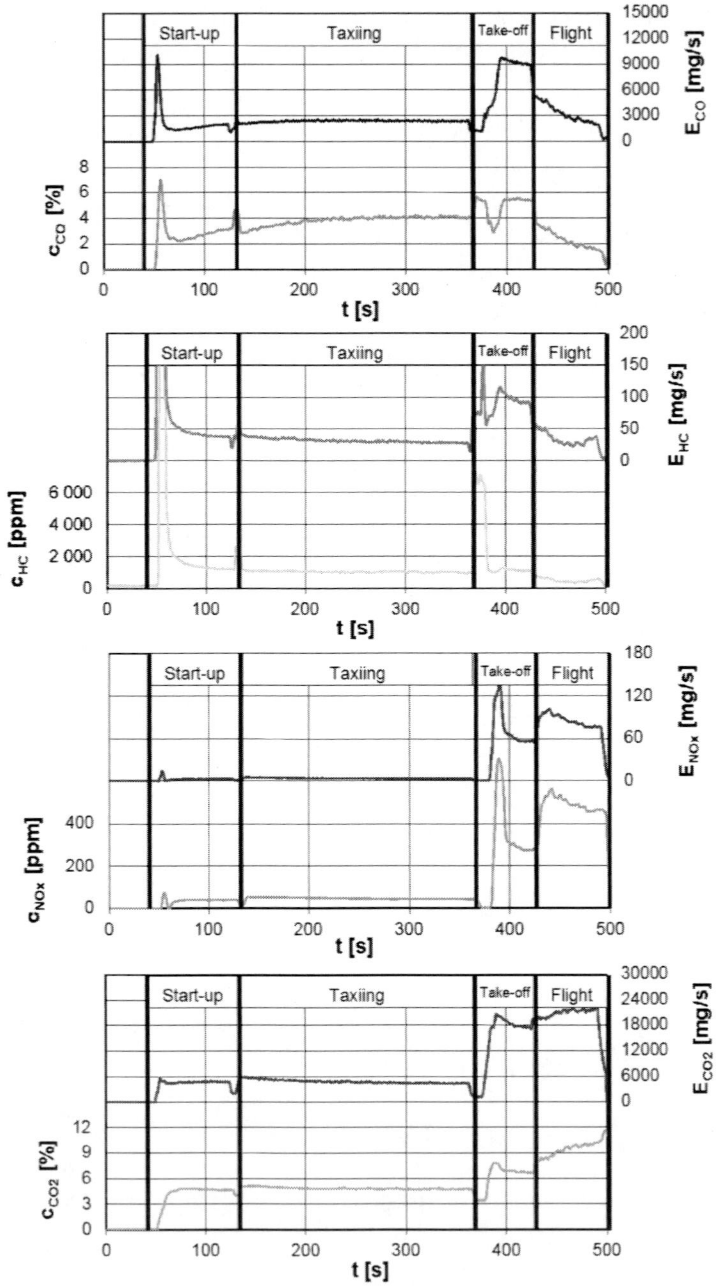

Figure 2.26. Changes in toxic compounds concentration and flow rate during a pre-flight test for Zlin-142 M.

Table 2.9. Measurement results of toxic compounds emission in a pre-flight test for Zlin-142 M

Parameter	Operation phase			
	start-up	taxiing	take-off	steady flight
Engine speed [rpm]	–	1200	2600	2400
Duration [s]	80	240	40	78
Mean concentration:				
c_{CO} [%]	3.20	4.21	5.54	2.57
c_{HC} [ppm]	3897	1144	1134	524
c_{NOx} [ppm]	26.6	34.0	276.9	337.2
c_{CO2} [%]	4.65	5.24	7.92	10.51
Mass [g]				
m_{CO}	169	560	329	258
m_{HC}	16.42	7.94	3.74	2.87
m_{NOx}	0.24	0.83	2.97	6.03
m_{CO2}	353	1115	1050	1539
Hourly emission [g/h]				
E_{CO}	7622	8405	29 575	11 923
E_{HC}	739	119	336	133
E_{NOx}	10.9	12.4	267.6	278.5
E_{CO2}	15 901	16 728	94 530	71 031

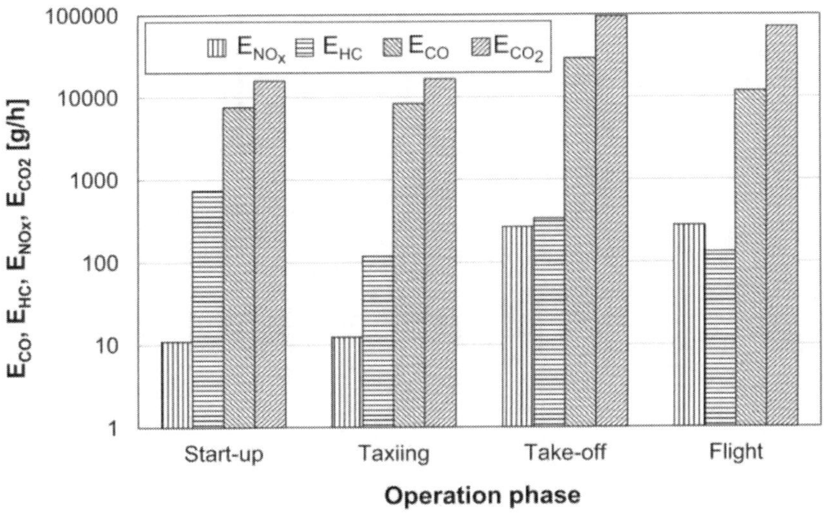

Figure 2.27. Mean hourly emissions of toxic compounds measured in a pre-flight test for Zlin-142 M.

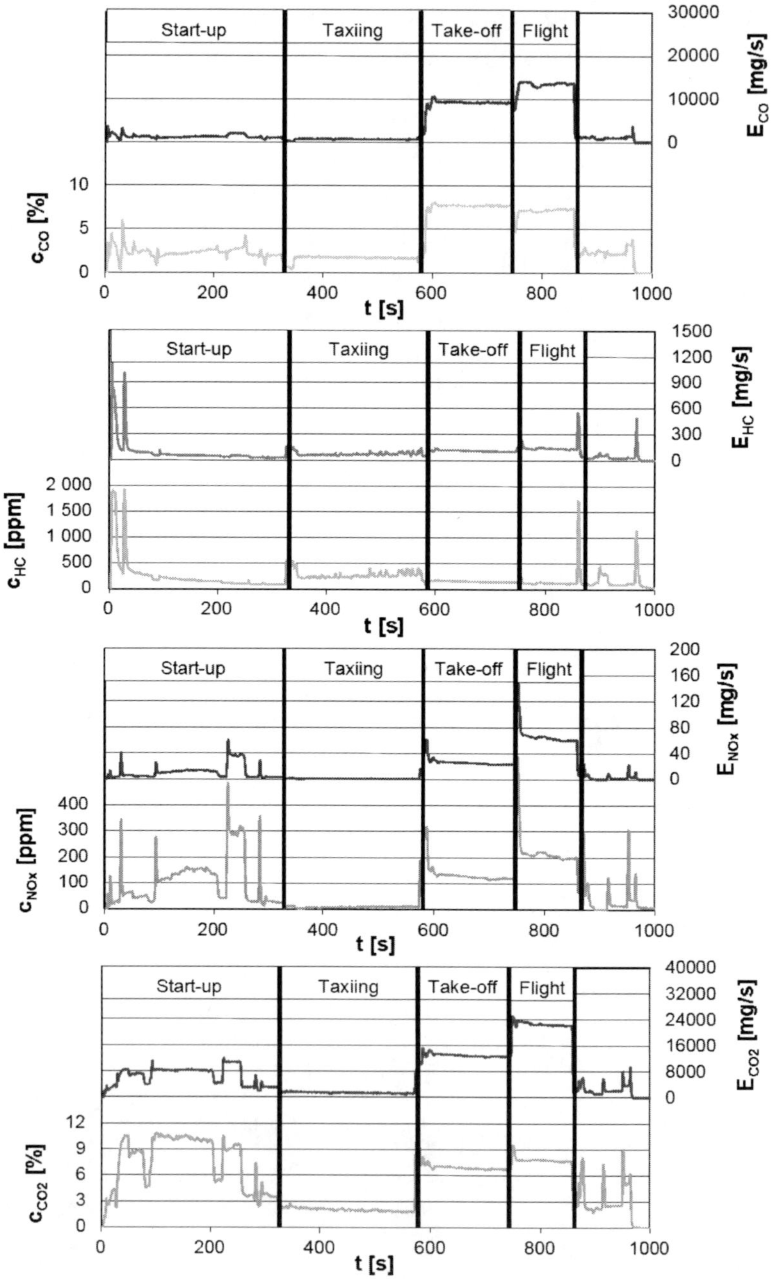

Figure 2.28. Changes in toxic compounds concentration and flow rate during a pre-flight test for Cirrus SR20.

Table 2.10. Measurement results of toxic compounds emission in a pre-flight test for Cirrus SR20

Parameter	Operation phase			
	start-up	taxiing	take-off	steady flight
Engine speed [rpm]	325	244	177	112
Duration [s]	–	1600	2800	2800
Mean concentration:				
c_{CO} [%]	2.54	1.72	8.41	8.37
c_{HC} [ppm]	512	558	321	262
c_{NOx} [ppm]	125	9.30	154	261
c_{CO2} [%]	8.30	2.11	8.13	9.20
Mass [g]				
m_{CO}	411	183	1491	1473
m_{HC}	28.8	18.2	18.9	16.0
m_{NOx}	3.35	0.14	4.40	7.56
m_{CO2}	2085	353	4416	2539
Hourly emission [g/h]				
E_{CO}	4554	2698	30 330	47 353
E_{HC}	319	269	384	515
E_{NOx}	37.1	2.12	89.5	243
E_{CO2}	23 095	5209	89 798	81 612

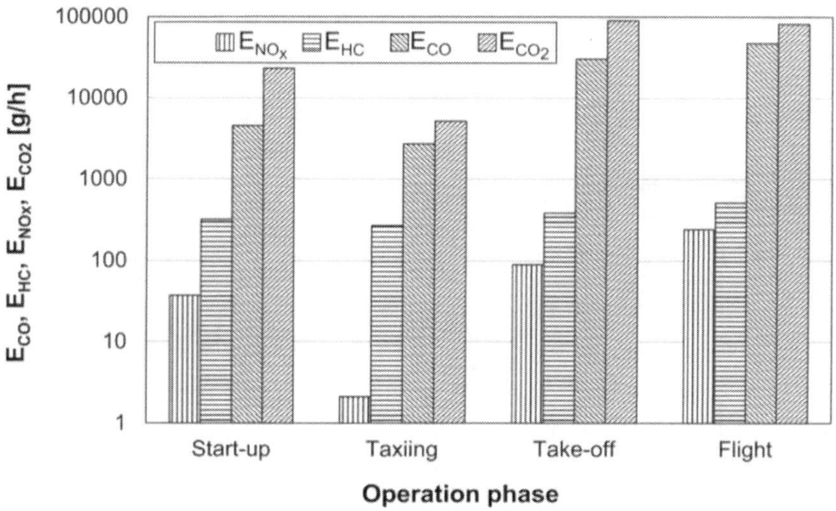

Figure 2.29. Mean hourly emissions of toxic compounds measured in a pre-flight test for Cirrus SR20.

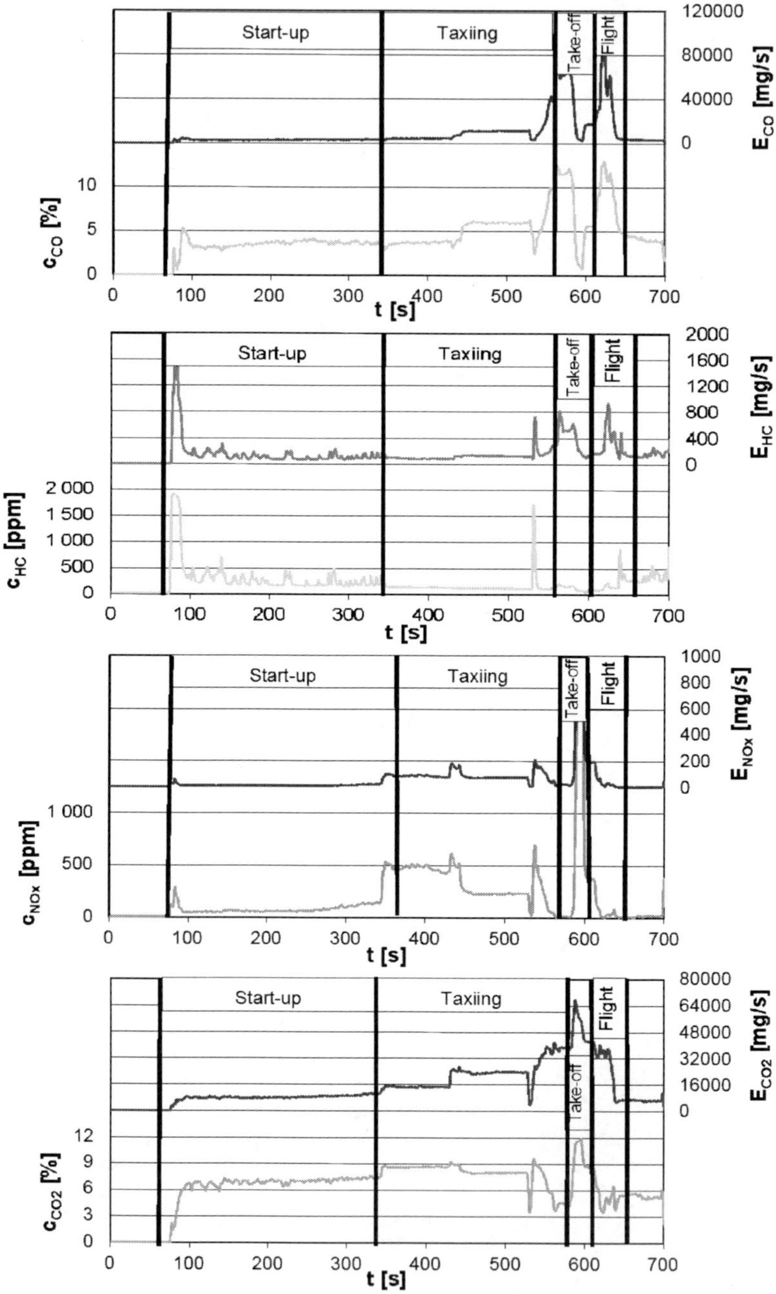

Figure 2.30. Changes in toxic compounds concentration and flow rate during a pre-flight test for SR22T (SP-AVB).

Table 2.11. Measurement results of toxic compounds emission in a pre-flight test for Cirrus SR22T (SP-AVB)

Parameter	Operation phase			
	start-up	taxiing	take-off	steady flight
Engine speed [rpm]	–	1600	2700	2700
Duration [s]	268	188	34	67
Mean concentration:				
c_{CO} [%]	3.92	5.41	5.69	5.38
c_{HC} [ppm]	669	289	174	650
c_{NOx} [ppm]	103	473	1112	32.7
c_{CO2} [%]	7.54	9.51	10.6	5.89
Mass [g]				
m_{CO}	748	1396	554	372
m_{HC}	45.6	21.8	6.04	11.9
m_{NOx}	3	15.8	15.4	0.31
m_{CO2}	2248	3636	1614	1653
Hourly emission [g/h]				
E_{CO}	10 059	26 667	58 644	20 015
E_{HC}	468	368	640	641
E_{NOx}	39.3	304	1626	16.9
E_{CO2}	30 204	69 624	170 920	88 873

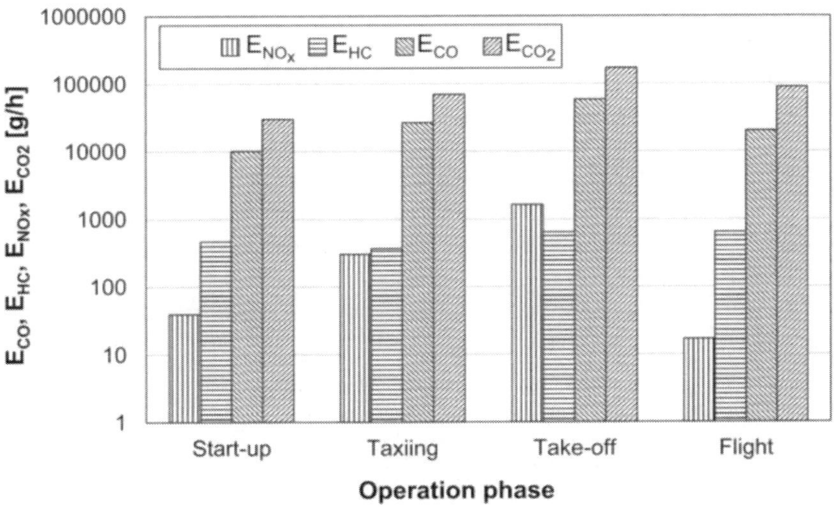

Figure 2.31. Mean hourly emissions of toxic compounds measured in a pre-flight test for Cirrus SR22T (SP-AVB).

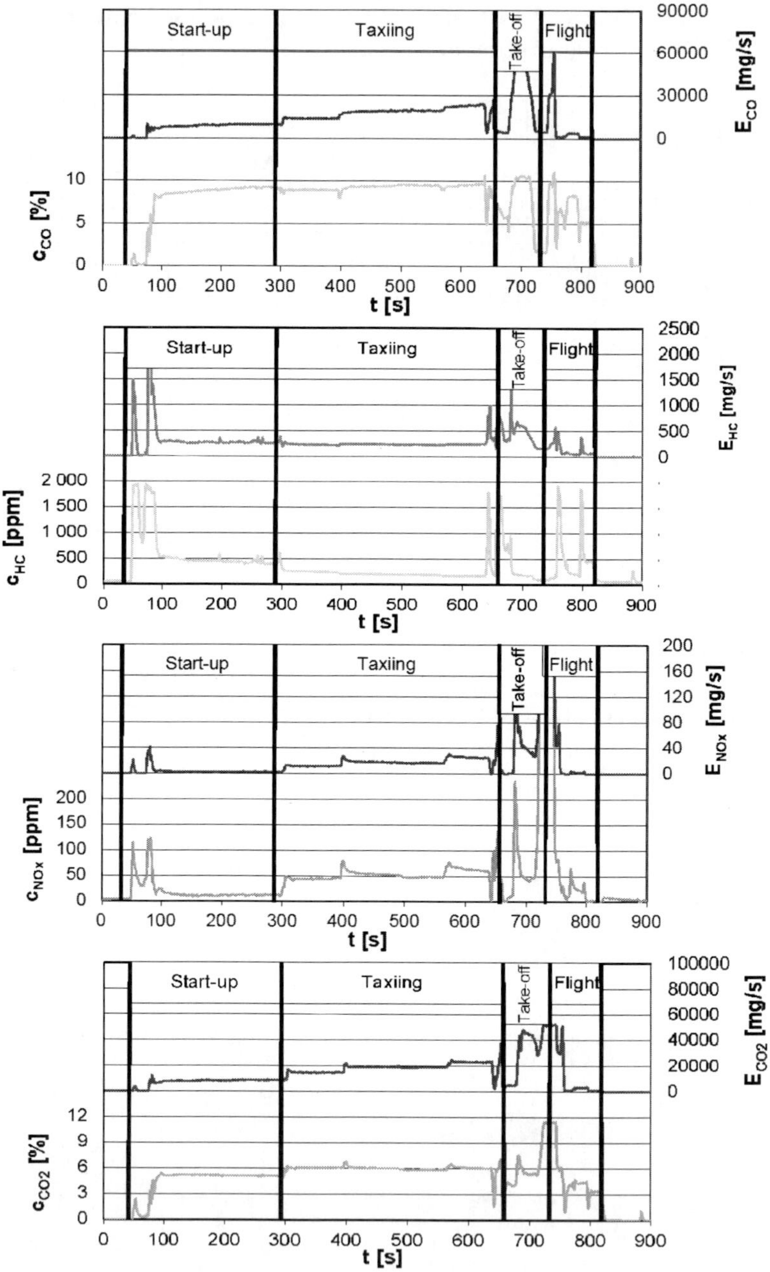

Figure 2.32. Changes in toxic compounds concentration and flow rate during a pre-flight test for SR22T (SP-AVE).

Table 2.12. Measurement results of toxic compounds emission in a pre-flight test for Cirrus SR22T (SP-AVE)

Parameter	Operation phase			
	start-up	taxiing	take-off	steady flight
Engine speed [rpm]	–	1600	2700	2700
Duration [s]	254	370	32	33
Mean concentration:				
c_{CO} [%]	9.12	10.7	4.1	8.2
c_{HC} [ppm]	1294	545	215.8	835.1
c_{NOx} [ppm]	25.9	67.3	1712.7	30.5
c_{CO2} [%]	5.32	6.91	11.7	4.8
Mass [g]				
m_{CO}	2004	6502	345.5	88.5
m_{HC}	90.8	97.1	6.14	2.34
m_{NOx}	0.83	6.35	20.2	15.1
m_{CO2}	2792	7960	1208	960.9
Hourly emission [g/h]				
E_{CO}	28 403	63 264	38 873	9651
E_{HC}	1287	945	691	256
E_{NOx}	11.7	61.8	2275.0	66.4
E_{CO2}	39 571	77 446	135 961	104 829

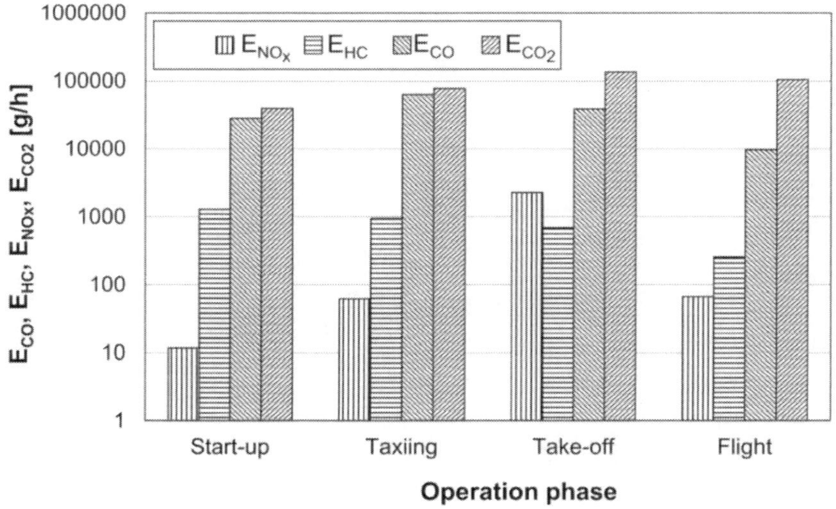

Figure 2.33. Mean hourly emissions of toxic compounds measured in a pre-flight test for Cirrus SR22T (SP-AVE).

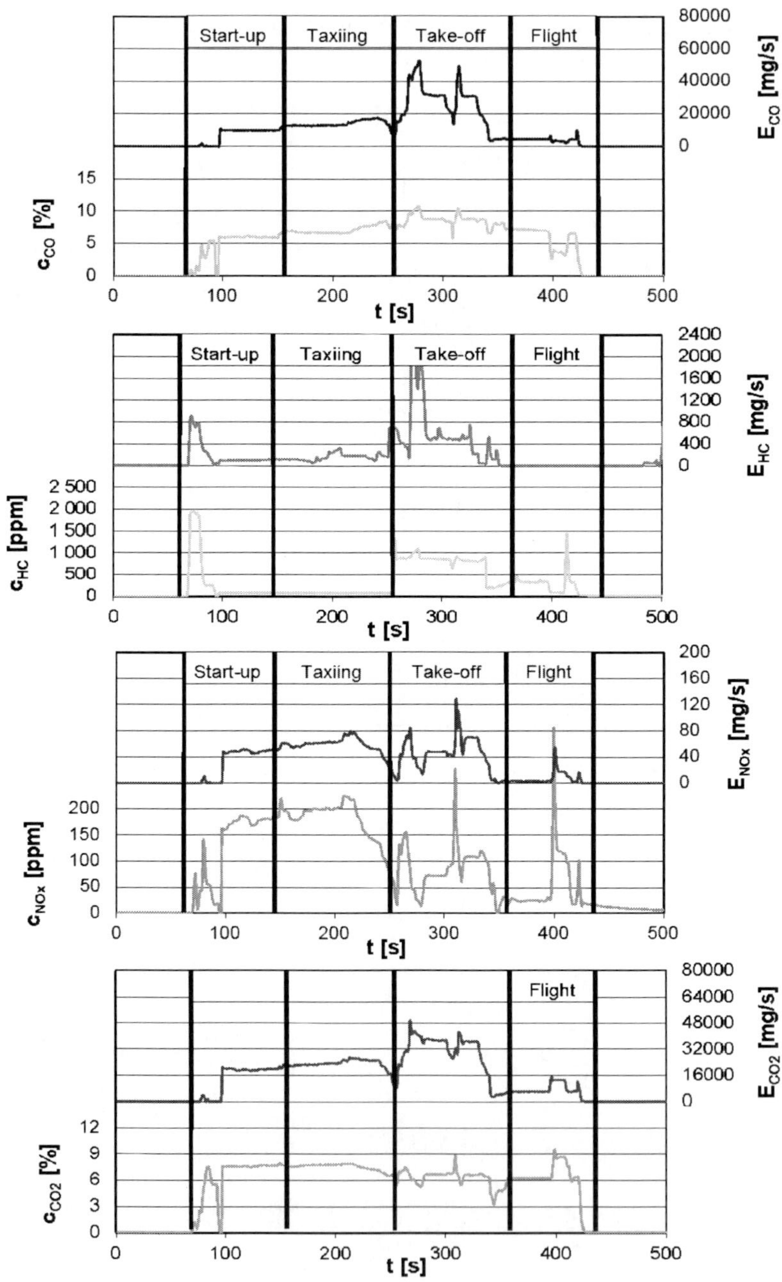

Figure 2.34. Changes in toxic compounds concentration and flow rate during a pre-flight test for Cirrus SR22T (SP-IES).

Table 2.13. Measurement results of toxic compounds emission in a pre-flight test for Cirrus SR22T (SP-IES)

Parameter	Operation phase			
	start-up	taxiing	take-off	steady flight
Engine speed [rpm]	–	1600	2700	2700
Duration [s]	82	95	98	81
Mean concentration:				
c_{CO} [%]	5.39	8.19	10.20	7.18
c_{HC} [ppm]	702	176	1532	610
c_{NOx} [ppm]	172	243	119	65.9
c_{CO2} [%]	7.01	8.87	7.60	7.12
Mass [g]				
m_{CO}	523	1316	2642	338
m_{HC}	9.69	9.81	15.8	14.8
m_{NOx}	2.59	5.80	4.58	0.52
m_{CO2}	1070	2241	2995	1805
Hourly emission [g/h]				
E_{CO}	22 943	49 876	97 070	15 004
E_{HC}	426	372	579	660
E_{NOx}	114	220	168	23.2
E_{CO2}	46 968	84 929	110 026	80 266

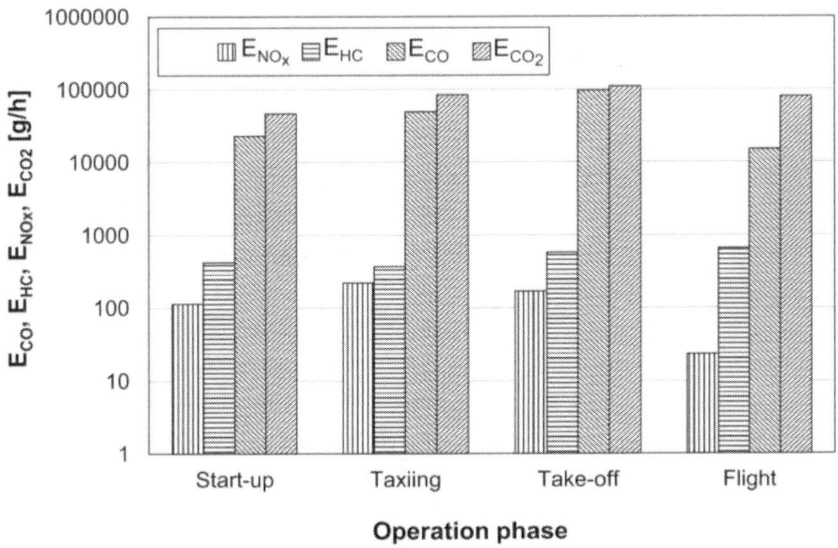

Figure 2.35. Mean hourly emissions of toxic compounds measured in a pre-flight test for Cirrus SR22T (SP-IES).

- Cessna 152 II – changes in toxic compounds concentration and flow rate during a pre-flight test (Figure 2.24), measurement results of toxic compounds emission in a pre-flight test (Table 2.8), results of mean hourly emissions of measured toxic compounds in a pre-flight test (Figure 2.25);
- Zlin-142 M – changes in toxic compounds concentration and flow rate during a pre-flight test (Figure 2.26), measurement results of toxic compounds emission in a pre-flight test (Table 2.8), results of mean hourly emissions of measured toxic compounds in a pre-flight test (Figure 2.27);
- Cirrus SR20 – changes in toxic compounds concentration and flow rate during a pre-flight test (Figure 2.28), measurement results of toxic compounds emission in a pre-flight test (Table 2.10), results of mean hourly emissions of measured toxic compounds in a pre-flight test (Figure 2.29).
- Cirrus SR22T (SP-AVB) – changes in toxic compounds concentration and flow rate during a pre-flight test (Figure 2.30), measurement results of toxic compounds emission in a pre-flight test (Table 2.11), results of mean hourly emissions of measured toxic compounds in a pre-flight test (Figure 2.31);
- Cirrus SR22T (SP-AVE) – changes in toxic compounds concentration and flow rate during a pre-flight test (Figure 2.32), measurement results of toxic compounds emission in a pre-flight test (Table 2.12), results of mean hourly emissions of measured toxic compounds in a pre-flight test (Figure 2.33);
- Cirrus SR22T (SP-IES) – changes in toxic compounds concentration and flow rate during a pre-flight test (Figure 2.34), measurement results of toxic compounds emission in a pre-flight test (Table 2.13), results of mean hourly emissions of measured toxic compounds in a pre-flight test (Figure 2.35);

Mean hourly emission values in each operation phase depend on the engine's capacity, current condition, design and the capacity-to-power ratio. For reasons presented above, those values are not readily comparable amongst one another. However, the resultant mean hourly emission of each individual compound ascribed to individual operation phases of the engine makes it possible to determine specific emissions. In order to do so it is necessary to take into account the mean engine load in a given operation phase. Mean engine loads for each operation phase were determined previously, on the basis of engines' operation parameters recorded in in-flight conditions (Chapter 2.2). The resultant load values were presented in relation to the effective power and then attributed to consecutive phases of the proposed test (Chapter 2.7). The next stage of the study consists of a test-based assessment of specific emissions of toxic compounds in exhaust gas, as described in the following chapter.

2.7. Test-Based Emission Assessment

The assessment of toxic emissions in exhaust gas from the selected airplane engines based on the proposed tests was carried out with reference to the measured concentrations of toxic compounds in exhaust gas recorded during pre-flight tests, as well as to mean hourly emission values calculated for each operation phase. The comparison of emissions from engines of differing designs is possible with the use of values correlated with the engine's power. Therefore, specific emission of each compound expressed in g/kWh is used as the unit of comparison. The resultant specific emission values for each compound corresponding to individual engine operation states were used to assess the proposed all-purpose aircraft emission tests. Start-up and taxiing, the two operation phases described in Chapter 2.3, belong to phase A, in accordance with the test's underlying assumptions. Therefore, the hourly emission values from those operation phases were averaged and the resultant value was ascribed to phase A. By reference to the characteristics of the engines used in the selected airplanes, engine power values in each test phase were determined. Subsequently, based on the power value corresponding to the phase in question and the hourly emission value for each compound, specific emission values were calculated. The results for each airplane are presented in Table 2.14.

The resultant specific emission values for each test phases were multiplied by the respective phase share coefficients in accordance with the following formula:

$$e_j = \frac{\sum E_{ji} \cdot u_i}{\sum N_{e,i} \cdot u_i} \quad [\text{g/kWh}] \tag{2.2}$$

where:

E_{ji} – hourly emission of the j-th pollutant of the i-th test phase i [g/h],

u_i – share coefficient of i-th phase of the test [–],

$N_{e,i}$ – engine power in the i-th phase of the test [kW];

thus obtaining specific emissions in the proposed test (Table 2.15).

Table 2.14. Mean emission rate and specific emission of toxic compounds from aviation engines

Airplane	Parameter	Test phase		
		A	B	C
PZL-104 Wilga 35	Engine power [kW]	39	194	136
	Hourly emission [g/h]			
	E_{CO}	20 968	53 054	30 515
	E_{HC}	1900	2470	1463
	E_{NOx}	115	544	266
	E_{CO2}	24 894	118 802	78 986
	Specific emission [g/kWh]			
	e_{CO}	540	273	225
	e_{HC}	49,0	12.7	10.8
	e_{NOx}	2.97	2.80	1.96
	e_{CO2}	642	612	582
Cessna 152 II	Engine power [kW]	17	86	60
	Hourly emission [g/h]			
	E_{CO}	3693	31 721	19 731
	E_{HC}	299	946	507
	E_{NOx}	49.8	283	133
	E_{CO2}	26 506	67 141	43 943
	Specific emission [g/kWh]			
	e_{CO}	215	369	328
	e_{HC}	17.4	11.0	8.43
	e_{NOx}	2.90	3.29	2.21
	e_{CO2}	1541	781	730
Zlin-142 M	Engine power [kW]	37	185	130
	Hourly emission [g/h]			
	E_{CO}	8013	29 575	11 923
	E_{HC}	429	336	133
	E_{NOx}	11.7	268	278
	E_{CO2}	16 315	94 530	71 031
	Specific emission [g/kWh]			
	e_{CO}	217	160	92.1
	e_{HC}	11.6	1,82	1.02
	e_{NOx}	0.32	1.45	2.15
	e_{CO2}	441	708	548

Airplane	Parameter	Test phase		
		A	B	C
Cirrus SR20	Engine power [kW]	29	145	102
	Hourly emission [g/h]			
	E_{CO}	3626	30 330	47 353
	E_{HC}	294	384	515
	E_{NOx}	20	90	243
	E_{CO2}	14 152	89 798	81 612
	Specific emission [g/kWh]			
	e_{CO}	125	209	467
	e_{HC}	10.1	2.65	5.08
	e_{NOx}	0.68	0.62	2.39
	e_{CO2}	488	620	804
Cirrus SR22T (SP-AVB)	Engine power [kW]	47	233	163
	Hourly emission [g/h]			
	E_{CO}	18 363	58 644	20 015
	E_{HC}	418	640	641
	E_{NOx}	172	1626	16.9
	E_{CO2}	49 914	170 920	88 873
	Specific emission [g/kWh]			
	e_{CO}	394	252	123
	e_{HC}	8.97	2.75	3,93
	e_{NOx}	3.69	6.98	0,10
	e_{CO2}	1071	734	545
Cirrus SR22T (SP-AVE)	Engine power [kW]	47	233	163
	Hourly emission [g/h]			
	E_{CO}	45 833	38 873	9651
	E_{HC}	1116	691	256
	E_{NOx}	37	2275	66
	E_{CO2}	58 509	135 961	104 829
	Specific emission [g/kWh]			
	e_{CO}	975	166.8	59.2
	e_{HC}	23.7	3.0	1.6
	e_{NOx}	0.81	9.8	0,4
	e_{CO2}	1345	583	643
Cirrus SR22T (SP-IES)	Engine power [kW]	47	233	163
	Hourly emission [g/h]			
	E_{CO}	36 409	97 070	15 004
	E_{HC}	399	579	660
	E_{NOx}	167	168	23
	E_{CO2}	65 949	110 026	80 265
	Specific emission [g/kWh]			
	e_{CO}	781	417	92.0
	e_{HC}	8.55	2.48	4.05
	e_{NOx}	3.58	0.72	0.14
	e_{CO2}	1415	472	493

Table 2.15. Specific values of toxic emissions for the selected airplanes in the proposed emission tests

Airplane	Test phase	A	B	C
	$N_e/N_{e\,max}$ [%]	20	100	70
	phase share	0.115	0.045	0.800
	specific emission determined in the test [g/kWh]			
	e_{CO}	e_{HC}	e_{NOx}	e_{CO2}
PZL-104 Wilga 35	239	12.3	2.05	585
Cessna 152 II	327	8.97	2.32	766
Zlin-142 M	101	1.47	2.02	540
Cirrus SR20	434	5.06	2.19	775
Cirrus SR22T (SP-AVB)	142	4.03	0.73	577
Cirrus SR22T (SP-AVE)	101	2.49	1.09	661
Cirrus SR22T (SP-IES)	141	4.10	0.31	524

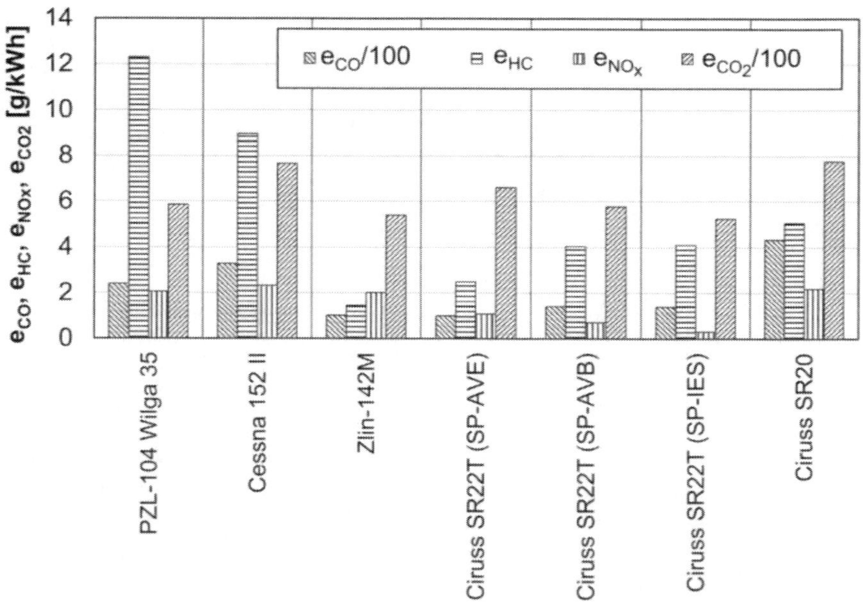

Figure 2.36. Comparison of test-derived specific emissions of pollutants for each aircraft.

Specific emissions of toxic compounds determined by means of the test were compared to one another (Figure 2.36). The distribution of specific emissions of pollutants from all tested propulsion systems is determined by the design of the engine and the construction of the airplane's propulsion system. Among all analyzed propulsion systems, there is a striking similarity of specific emissions from Cirrus SR22T airplanes. All planes in this group have the same structural parameters and use the same propulsion system: the same type of variable pitch airscrew and the same engine type. None of those airplanes was in operation for more than 5 years before the test and their respective technical conditions were similar.

While Cirrus SR20 is also similar to the Cirrus SR22T family in terms of design, it differs in terms of the engine and is fitted with an earlier version of avionics. The engine in Cirrus SR20 is naturally aspirated and of similar design, but has a smaller capacity and manual control of the fuel and air mixture composition. The engines of those airplanes are produced by the same manufacturer using a similar construction technology. The next group of propulsions systems generating similar specific emissions are those used in PZL-104 Wilga 35 and Zlin-142 M. These planes are propelled by a fixed pitch airscrew, in both cases powered by an engine supercharged with a radial compressor. Both of these airplanes had been in operation for over 20 years before the tests. One airplane that stands out from all others is Cessna 152 II, fitted with a fixed pitch airscrew powered with a naturally aspirated engine. In its case, the operation period had also been greater than 20 years. On the basis of the airplanes' characteristic features and their specific emission values, mean emissions of pollutants were determined (Table 2.16). The mean emission values for each of the pollutant were calculated on the basis of values measured in Cirrus propulsion systems, as Cirrus planes were considered to be representative of state-of-the-art aviation solutions.

Table 2.16. Specific emissions of pollutants for Cirrus airplanes and their mean values in the proposed tests

Airplane	**Cirrus SR22T (SP-AVB)**	**Cirrus SR22T (SP-AVE)**	**Cirrus SR22T (SP-IES)**	**Cirrus SR20**	**Mean value**
Pollutant	**test-derived specific emission [g/kWh]**				
CO	142	101	141	434	**205**
HC	4.03	2.36	4.10	5.09	**3.9**
NO_x	0.73	1.10	0.31	2.20	**1.1**
CO_2	578	661	524	775	**635**

An analysis of percentage differences between specific emissions from all tested airplanes on the one hand and mean specific emissions determined on the basis of emissions from Cirrus airplanes on the other (Table 2.17 and Figure 2.37) indicates changes resulting from the design of the propulsion system and the current technical condition of airplanes.

The engine fuel supply system is one of the elements affecting the emission of pollutants in the analyzed propulsion systems. In the cases of PZL-104 Wilga 35, Cessna 152 II and Zlin-142 M the fuel supply system is fitted with a carburetor, whose performance determines the operational and environmental performance of the engine as a whole. Engines fitted with carburetor fuel supply systems generate greater specific emissions of hydrocarbons; in this case the structural design of the engine is of even greater importance. The cylinders in the engine of the PZL-104 Wilga 35 are arranged radially, whereas the engine of Zlin-142 M has a straight cylinder arrangement with downward oriented cylinders. In the case of an excessive wear of the piston-and-cylinder assembly, such a structure also causes a greater concentration (and thus specific emission) of hydrocarbons, because engine oil penetrates into the engine's exhaust system. In the case of the PZL-104 Wilga 35 the engine service life was nearly over, causing poorer technical condition translating into significantly greater specific emission of hydrocarbons (approximately 280% above the mean value) and simultaneously into a relatively small percentage difference in the emissions of carbon monoxide and carbon dioxide, caused by unburnt mass of fuel introduced into the cylinder.

Table 2.17. Percentage differences in specific emissions compared to mean specific emissions from Cirrus airplanes

Airplane	PZL-104 Wilga 35	Cessna 152 II	Zlin-142 M	Cirrus SR20	Cirrus SR22T (SP-AVE)	Cirrus SR22T (SP-AVB)	Cirrus SR22T (SP-IES)
Pollutant	Difference in specific emissions of pollutants compared to mean specific emissions from Cirrus airplanes [%]						
CO	17	60	–50	113	–51	–31	–31
HC	214	130	–63	29	–37	3	5
NO_x	90	115	87	103	1	–33	–71
CO_2	–8	21	-15	22	4	–9	–17

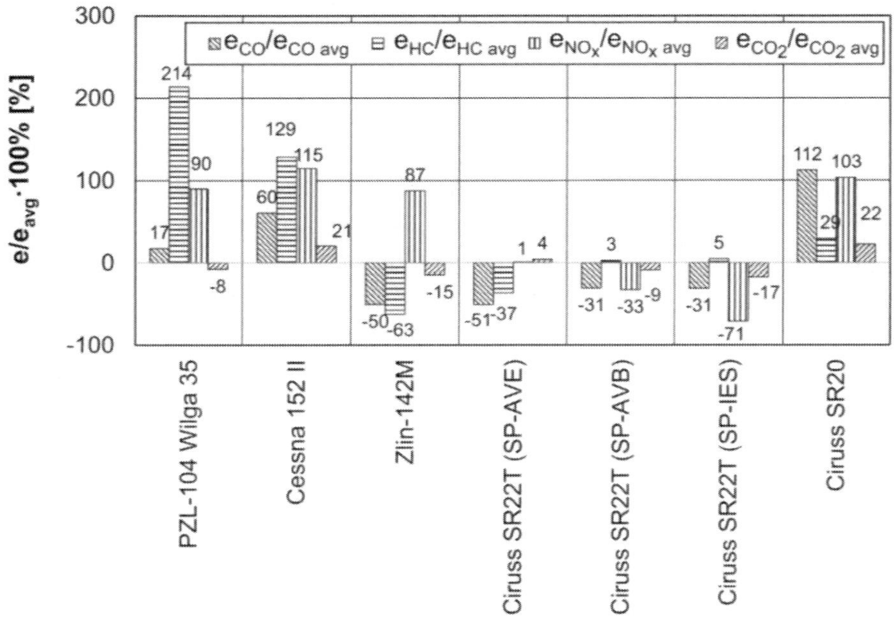

Figure 2.37. Percentage differences in specific emissions of each compound against the mean emissions from Cirrus airplanes.

Attention should be drawn (Figure 2.37) to increased specific emissions from naturally aspirated engines used in the propulsion systems of Cessna 152 II and Cirrus SR20. Low engine capacity-to-power ratio as shown in the equation 3.2 translates into greater specific emissions of pollutants. The beneficial effect of engine supercharging is visible in the case of propulsion systems used Cirrus SR22T and Zlin-142 M, where the differences in specific emissions usually do not exceed 70%. A comparison of specific emissions of individual compounds for these airplanes additionally highlights the role of the fuel supply system parameters. A high degree of similarity in specific emissions is observed for all Cirrus SR22T airplanes, which is owed to the use of the fuel dose adjustment system that analyses external parameters (pressure and temperature) coupled with information on the throttle position. In the case of Zlin-142 M the settings of the fuel supply and control systems depend on the experience of the operator carrying out the test. In such a case the specific emissions of carbon monoxide and hydrocarbons amount to a half of the corresponding mean value. In their turn, the specific emissions of nitrogen oxides and carbon dioxide exceed the corresponding mean values by 100%

and 50%, respectively. The growth in these emissions results from the fact that the excess air coefficient in the engine is close to one.

Comparison of specific carbon dioxide emissions measured in accordance with the proposed test procedure against mean carbon dioxide emissions in the group of supercharged and naturally aspirated engines indicates that the differences between those emissions and the corresponding mean values are similar, which allows one to conclude that the assumed test phase coefficients are correct in terms of similarity of energy demand of the tested engines.

An analysis of percentage differences between specific emissions of each pollutant and the corresponding mean emissions from newer airplanes determined via the proposed test procedure shows that it is possible to use that test procedure to evaluate the airplanes' technical condition.

It needs to be stressed that the emission test in accordance with the proposed procedure can be carried out for the purpose of assessing the emission of pollutants and the technical condition of airplanes. However, to be able to obtain information on operational and environmental performance of aircraft propulsion systems it is necessary to determine the allowed specific emission limits for each pollutant contained in exhaust gas, approximately 15% above the determined mean values. It is proposed that the allowed specific emission limits be as follows:

$$e_{CO} = 235 \text{ g/kWh},$$
$$e_{HC} = 4.6 \text{ g/kWh},$$
$$e_{NO_x} = 1.15 \text{ g/kWh}.$$

Emissions from aviation piston engines exceeding the limits proposed above could provide a stimulus for verifying the technical condition of the engine.

The Authors' attempt at assessing specific emissions of pollutants in exhaust gas from piston engines used to power small aircraft is aimed at determining the environmental impact of general aviation aircraft. Therefore, it requires an assessment of the actual operating conditions of those engines, as well as measurements of specific emissions in such conditions. To do so, the specificity of their operation must be addressed. On the basis of the Authors' publications, studies and analyses one can conclude that the proposed simplified stationary test for general aviation reflects – to some extent – actual in-flight conditions and can therefore be used in studies involving a larger number of aircraft in order to determine their environmental impact. The said procedures can be accelerated by simplification of procedures (e.g., the use of emission models for aviation piston engines). The following chapter proposes the methodology for modeling the emission (or mass) of pollutants generated by aircraft in in-flight conditions.

2.8. MODELING OF EXHAUST EMISSIONS

2.8.1. Preliminary Assumptions

Contemporary studies into propulsion systems involving the search for relationships between different characteristics increasingly rely on the experiment planning theory, due to the high number of such characteristics [1, 6, 18, 23, 29, 38–40, 46, 60]. Therefore, the construction of the mathematical model began with an assumption on its structure, followed by identification of parameters. Subsequent analysis of the model made it possible to formulate conclusions with a degree of generality higher than at the stage of interpreting measurement results. Pollutants' emission models were identified and their characteristics were determined based on the input and output values.

The constructed emission models are tailored to general aviation propulsion systems. The Authors' studies and analyses of aircraft propulsion show that it is necessary to develop models for two propulsion systems powered by aviation piston engines. The first of those model groups is dedicated to fixed pitch airscrew systems, while the other is dedicated to variable pitch airscrew systems. The emission models were built using the results of measurements obtained in stationary tests for PZL-104 Wilga 35 and Cirrus SR22T (SP-IES) airplanes.

To complete the above task it was necessary to measure the engine's operational parameters, as well as exhaust gas parameters – concentration and flow rate of gaseous compounds. The next determined value was the input values vector X (Figure 2.38) affecting the output vector **Y** – the mass of gaseous compounds contained in exhaust gas.

The constant values vector C and its possible effect on emissions was taken into account. These values did not change during the experiment (the said effect resulted e.g., from the type of the lubricating oil used, etc.). The model also features the interfering values vector Z, i.e., values that were subject to change (e.g., temperature, pressure and relative humidity of ambient air, as well as other values, not explicitly presented in the model).

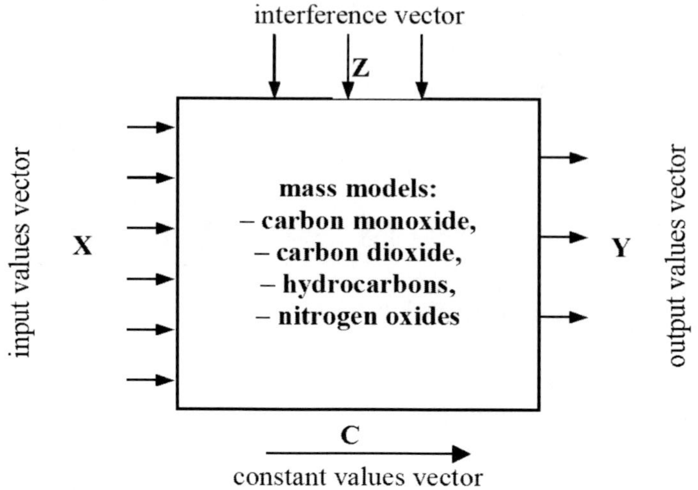

Figure 2.38. Characteristic values addressed in the experiment.

The model's input values vector X consists of the following:

x_1 engine speed, n [rpm],
x_2 exhaust gas temperature, T_{exh} [K],
x_3 excess air coefficient, λ [–],
x_4 test-derived specific emission of carbon monoxide, e_{CO} [g/kWh],
x_5 test-derived specific emission of carbon dioxide, e_{CO2} [g/kWh],
x_6 test-derived specific emission of hydrocarbons, e_{HC} [g/kWh],
x_7 test-derived specific emission of nitrogen oxides, e_{NOx} [g/kWh],
x_8 maximum effective engine power, $N_{e\,max}$ [kW],
x_9 flight altitude, h_{fly} [m],
x_{10} ascending velocity, v_{asc} [m/s],
x_{11} descending velocity, v_{dsc} [m/s],
x_{12} flight duration, t_{fly} [s],
x_{13} duration of ground operations, t_{op} [s].

Output values vector **Y** consists of the following:

y_1 mass of carbon monoxide, m_{CO} [g],
y_2 mass of carbon dioxide, m_{CO2} [g],
y_3 mass of hydrocarbons, m_{HC} [g],
y_4 mass of nitrogen oxides, m_{NOx} [g].

Interfering values vector Z consists of the following:

z_1 air temperature, T_{air} [K],
z_2 air pressure, p [Pa],
z_3 relative humidity, H [%].

Constant values vector **C** consists of the following:

c_1 engine capacity, V_{eng} [dm^3],
c_2 airscrew type (fixed pitch/variable pitch).

If tests are carried out in conditions similar to ambient conditions it is acceptable to ignore the interfering values. The emission of toxic substances from internal combustion engines is a process that particularly strongly depends on the engine's operation conditions as well as on many other factors: design, operation, external conditions, etc. As regards the tests in question, the appropriateness of the decision to ignore the interfering values was ensured by verifying the measurements in comparable conditions.

The values that characterize the operating conditions of the engine were determined on the basis of real-time readings of the on-board flight recorder (in the case of Cirrus SR22T) and by means of the data acquisition system (in the case of PZL-104 Wilga 35). Pollution parameters – concentration of toxic compounds – were measured by means of instruments compliant with the applicable standard.

Identification was made with regard to the model structure (structural identification) as well as its parameters (parametric identification). The identification of processes in machine design, usually determined by means of linear equations, frequently proves rather unrealistic. Product-form equations are often used in machine construction as they enable an uncomplicated analysis, which is why this type of equations (model) was assumed by the Authors and is used in further deliberations

$$y = x_1 \cdot \ldots \cdot x_n \qquad (2.3)$$

The assumed structural identification results from earlier studies and analyses into the relationships between emissions of toxic compounds in exhaust gas as well as the thermodynamic relationships taking place within internal combustion engines, expressed by means of product-form or power-form models. Therefore, when searching for the model for toxic emissions in exhaust gas from aviation piston engines, product-form models were assumed in accordance with the function similarity principle.

2.8.2. Pollutants Emission Models

Identification was made with regard to the inputs into models used for calculating the mass of pollutants emitted by aircraft. On that basis it was determined that of key importance are the values of specific emissions of each compound (e_j) obtained in the proposed test, as well as engine power (N_e) and flight duration (t_{fly}).

$$m_j = e_j \cdot N_e \cdot t_{fly} \tag{2.4}$$

where:

- m_j mass of the exhaust gas compound j [g],
- e_j specific emission of the exhaust gas compound j [g/kWh],
- N_e effective power resulting from a given load [kW],
- t_{fly} flight duration at a given load [h].

Therefore, the model needs to include such values as the duration of each flight phase, flight altitude, ascending velocity upon take-off and descending velocity upon landing. The duration of each operation phase is of key importance for mass modeling because together with engine operation parameters (engine speed and load) they directly affect the concentration of pollutants in exhaust gas and the mass of exhaust gas ejected from the engine. Therefore, the modeling exercise should begin with searching for a method of identifying the relative share of flight phases in the entire flight duration. When identifying flight phases one can assume that the flight duration t_{fly} is the sum of take-off and ascent times t_{asc}, steady flight time t_{steady}, and descent and landing time t_{land}, as shown by the following equation:

$$t_{fly} = t_{asc} + t_{steady} + t_{land} \text{ [s]} \tag{2.5}$$

Take-off and landing times depend on the airplane's performance, as well as on the airport's procedures for ascent after take-off and approaching to land. In commercial airports and for operational flights, these times are typically shorter than one hundred seconds and do not represent a major share in the entire flight duration. In the case of training flights, usually being shorter and involving smaller altitudes, the absolute durations of those flights are similar, but their share in the overall flight duration is significantly greater. Therefore, it is necessary to develop pollution mass models featuring a division of the entire flight duration into individual phases.

The duration of take-off time depends on the airplane's design and performance, i.e., the design of the airframe determining the minimum take-off velocity and the design of propulsion system, determining the airplane's capability to generate thrust necessary to gain the velocity required for take-off. The above structural and propulsion parameters are indirectly included in the parameters that describe any aircraft, namely ascending velocity (v_{acs}) and descending velocity (v_{dsc}). On the basis of those parameters (and the flight altitude) one can determine the duration of the take-off and ascent phase. Therefore, the equation (2.6) takes the following form:

$$t_{fly} = \frac{h_{fly}}{v_{asc}} + t_{steady} + \frac{h_{fly}}{v_{dsc}} \text{ [s]} \tag{2.6}$$

On the basis of flight altitude, ascent velocity and descent velocity (separately for each airplane) as well as the entire flight duration, the steady flight phase duration was determined. By transforming equation 3.5, the following equation (2.7) was arrived at:

$$t_{steady} = t_{fly} - h_{fly}\left(\frac{1}{v_{asc}} + \frac{1}{v_{dsc}}\right) \text{ [s]} \tag{2.7}$$

The method presented above was used to individualize model input parameters linked to the duration of each flight phase. As regards airplane operation, the following phases were identified as well: start-up, warming up, cooling down and taxiing to the runway or to the hardstand. Therefore, pollution mass models must necessarily address the existence of these phases, which was ensured by way of introducing the ground operation time parameter (t_{op}), with all of them collectively included.

An analysis of model input data showed that such parameters as engine speed, exhaust gas temperature and excess air coefficient had already been addressed at the stage of determining specific emissions of each compound. A differentiation of the propeller (airscrew) and the degree of engine capacity-to-power ratio are not directly stated in the model, but are addressed indirectly in the maximum effective engine power, ascent velocity and flight altitude. The power generated by the propulsion system in each flight phase was replaced with the maximum effective engine power. Therefore, the component corresponding to steady flight was multiplied by 0.7, while the component corresponding to ground operations was multiplied by 0.2.

By comparing the selected input parameters the following approximation functions (models) for the mass of toxic compounds in exhaust gas from aviation piston engines were identified:

- mass of carbon monoxide [g]:

$$m_{\mathrm{CO}} = \frac{e_{\mathrm{CO}} \cdot N_{e\,\max} \cdot W}{3600} \tag{2.8}$$

- mass of hydrocarbons [g]:

$$m_{\mathrm{HC}} = \frac{e_{\mathrm{HC}} \cdot N_{e\,\max} \cdot W}{3600} \tag{2.9}$$

- mass of nitrogen oxides [g]:

$$m_{\mathrm{NOx}} = \frac{e_{\mathrm{NOx}} \cdot N_{e\,\max} \cdot W}{3600} \tag{2.10}$$

- mass of carbon dioxide [g]:

$$m_{\mathrm{CO_2}} = \frac{e_{\mathrm{CO_2}} \cdot N_{e\,\max} \cdot W}{3600} \tag{2.11}$$

where:

$$W = \frac{h_{fly}}{v_{asc}} + 0.7 \cdot \left(t_{fly} - h_{fly} \cdot \left(\frac{1}{v_{asc}} + \frac{1}{v_{dsc}} \right) \right) + 0.2 \cdot \left(\frac{h_{fly}}{v_{dsc}} + t_{op} \right)$$

By generalizing the above formulae, one can develop a formula demonstrating the mass of a toxic compound emitted in exhaust gas during aviation operations:

$$m_j = \frac{e_j \cdot N_{e\,\max} \cdot W}{3600} \tag{2.12}$$

For demonstration purposes, the above models were used for calculating the mass of toxic compounds from aviation propulsion systems tested in the proposed stationary test (Table 2.18). For all airplanes, the same flight parameters were assumed (flight altitude h_{fly} of 500 m and flight duration t_{fly} of 3600 s). An analysis of operation parameters of general aviation aircraft made it possible to specify a unified time of airport operations, including engine start-up, warming up, cooling down and taxiing (t_{op} = 800 s). The maximum effective engine power $N_{e\ \max}$ was assumed on the basis of each engine's characteristics. Ascent velocity during take-off and descent velocity during landing were assumed on the basis of technical specifications of the airplanes. The values of specific emissions of individual compounds, considered as model input parameters, were determined on the basis of measurements carried out during the proposed emission test.

Table 2.18. Output values for mass models for toxic compounds in exhaust gas for the selected airplanes

Parameter	PZL-104 Wilga 35	Cessna 152 II	Zlin-142 M	Cirrus SR22T (SP-IES)	Cirrus SR20
			Input values		
$N_{e\,\max}$ [kW]	194	86	185	233	145
t_{fly} [s]			3600		
h_{fly} [m]			500		
v_{asc} [m/s]	5.0	3.6	4.4	6.0	4.2
v_{dsc} [m/s]	3.0	3.0	2.7	3.0	3.0
e_{CO} [g/kWh]	239	327	101	141	434
e_{HC} [g/kWh]	12.3	8.97	1.47	4.10	5.06
e_{NOx} [g/kWh]	2.05	2.32	2.02	0.31	2.19
e_{CO2} [g/kWh]	276	766	540	524	775
t_{op} [s]			800		
			Pollutant mass values derived from the models		
m_{CO} [g]	34 042	20 582	13 672	23 840	46 184
m_{HC} [g]	2119	563	198	696	540
m_{NOx} [g]	255	146	274	53	234
m_{CO2} [g]	82 773	48 108	72 695	88 935	82 177

The subsequent stage of the study consisted of verifying the above theoretical models against results of in-flight tests of toxic emissions from aircraft propulsion systems.

The results obtained on the basis of the proposed pollution mass models were assessed on the basis of measurements of toxic compounds taken in in-flight conditions for the Cirrus SR20 airplane. The Authors' experience in aircraft testing [30–37], new possibilities as regards ensuring state-of-the-art test units as well as new instrumentation capabilities all contributed to developing a methodology of in-flight testing of toxic emissions in exhaust gas from aviation piston engines.

The purpose of the study was to determine the mass of pollutants in exhaust gas generated in actual operating conditions. The said mass will constitute output data for comparative analysis of the results obtained by means of the models presented above. The airplane in question is a state-of-the-art unit; the airframe is made of a composite material and the propulsion system is fitted with a variable pitch airscrew powered with a naturally aspirated 6-cylinder piston engine (opposed-piston design). The airplane is fitted with modern avionics systems and an on-board recorder of flight parameters and key engine operation parameters. Another advantage is that the composition of the air and fuel mixture is not regulated automatically. The specific features of the propulsion system in the selected airplane are closely related to propulsion systems used in older generation aircraft, and avionics systems installed on the airplane make it possible to obtain additional operational information in a digital form. The cargo space and maximum load capacity allow an installation of measurement equipment. Such a combination of design features does not fully reflect the characteristics of all airplanes used in stationary (pre-flight) tests, but includes the highest number of such features. Therefore, the verification of mass models was carried out by means of the Cirrus SR20 airplane, and its results were used to develop a general emission model for airplanes powered by piston engines.

The in-flight tests were carried out by means of the same toxic compounds concentration measurement equipment (Semtech DS analyzer) that was used in stationary (pre-flight) tests. The exhaust gas analyzer and the necessary elements of the measurement system were installed on the airplane in a manner ensuring correct position of the center of gravity, thus recreating actual operation conditions. The exhaust gas analyzer was placed in the cargo space (Figure 2.39), and the battery providing a separate power source for the measurement equipment was placed behind the front passenger seat. For the purpose of measuring the emissions of toxic compounds in exhaust gas, the exhaust system was extended by 2.5 meters. As a result, the exhaust gas sampling point was located in a place enabling correct installation of the measurement probe (Figure 2.40). Such significant distance also served the purpose of reducing exhaust gas temperature – since the original exhaust system is very short, the temperature immediately downstream of the engine is approximately 1300 K, while the maximum temperature tolerated by the measurement equipment should not exceed 973 K. In order to minimize the interference with the engine's exhaust system, which may potentially reduce the engine's performance, measurements were made in one exhaust manifold.

Figure 2.39. Semtech DS exhaust gas analyzer on board of Cirrus SR20.

Figure 2.40. Location of the exhaust gas sampling probe.

The test flight originated from the Poznań-Ławica airport. The plane flew towards the town of Szamotuły where it turned back and landed on the airport of origin. The selected air space range in the controlled space of the Poznań-Ławica airport is rarely used for operational flights, and therefore the planned route did not conflict with the routes of other aircraft. The test operation involved engine start-up, warming up, taxiing to the runway, take-off and ascent to the desired cruising altitude, followed by steady flight with fuel and air mixture optimization, and subsequently by descent, approaching to land, landing and taxiing to the hardstand. As recommended for the flight in question originating from a commercial airport such as Poznań-Ławica, after the take-off in the eastern direction the plane took a north-western turn in the direction of Szamotuły. The flight was completed as planned. Toxic compounds in exhaust gas were successfully measured, and so were the operational parameters of the airplane and its propulsion system. The recorded environmental and operational performance parameters included the concentration of toxic compounds (carbon monoxide, hydrocarbons, nitrogen oxides and carbon dioxide) in exhaust gas and the exhaust gas flow rate, as well as engine speed, flight altitude and geographical latitude and longitude. The recorded GPS signal served as a basis for visualizing the airplane's flight trajectory (Figure 2.41).

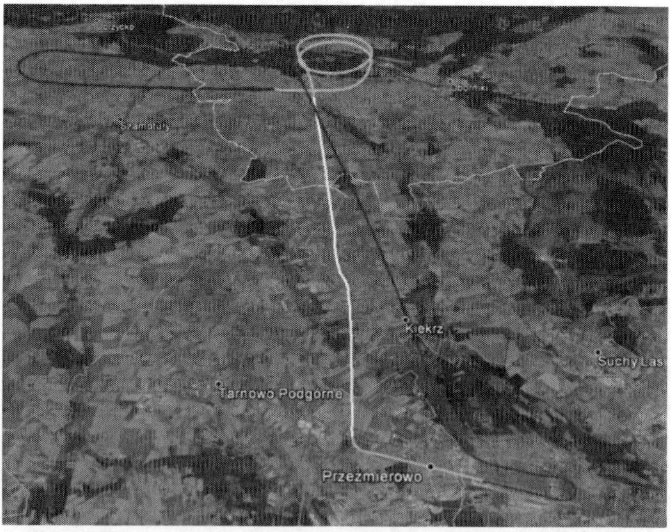

Figure 2.41. Test flight trajectory; taxi to take-off (blue), take-off and ascent (red), steady flight (magenta), descent (green), return to approach to land (yellow), approach to land and landing (orange), taxi to hangar (light blue).

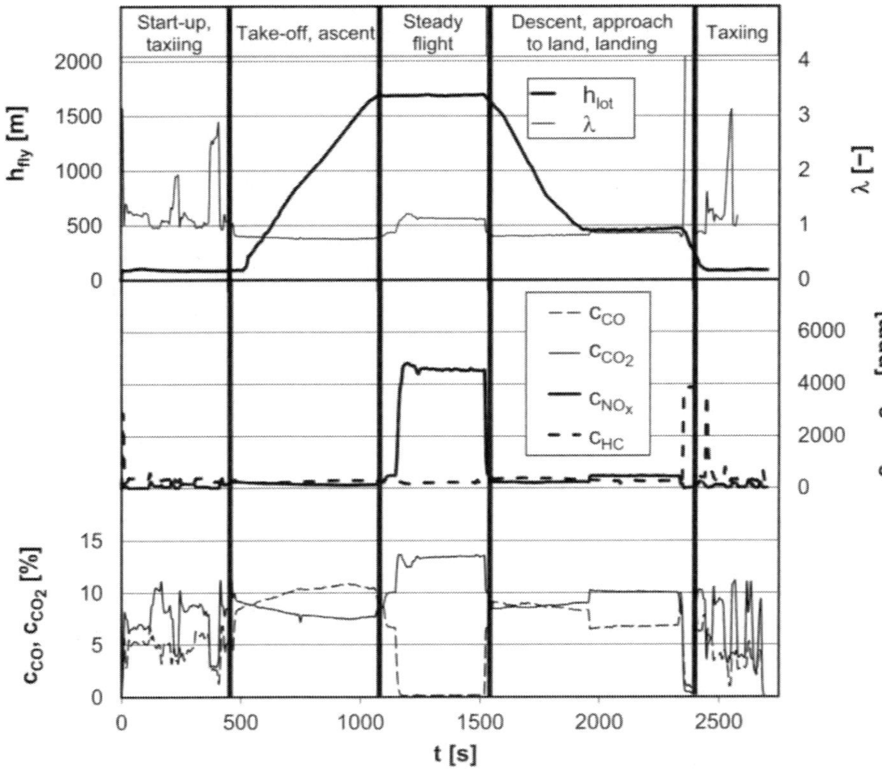

Figure 2.42. Concentration of toxic compounds in exhaust gas in each flight phase.

The engine's operating conditions are closely related to the nature of changes in flight altitude and are reflected in the concentration of toxic compounds in exhaust gas (Figure 2.42). The function plots presented below reflect the nature of changes in the concentration of each pollutant, connected with operating conditions resulting from changes in flight altitude and the possibility to adjust the composition of the air and fuel mixture. The greatest concentration changes caused by changes in the engine's operating parameters can be seen in the cases of carbon monoxide and nitrogen oxides. Specifically, these changes took place in the steady flight phase upon adjustment of the air and fuel mixture composition (Figure 2.43).

The test results denote a close correlation of the engine's performance parameters with changes in the concentrations and emission rates of toxic compounds in exhaust gas. If engine speed is constant, the above changes depend on the airscrew pitch and flight parameters, each of which affects the engine load. For start-up, warming up and taxiing to the runway and then to the hardstand, the engine operation conditions are characteristic for high engine speed variability and minor changes in engine load, which is reflected in changing rates of toxic emissions. In the other flight phases, the engine and airplane performance values were fairly constant. This means that the take-off ascent had a constant speed at maximum fuel supply to the engine and at maximum engine power. Concentrations of toxic compounds in exhaust gas in the take-off and ascent phases had the following values: from the beginning of the take-off the carbon dioxide concentration amounted to approximately 10% and then gradually decreased as the altitude grew. After the desired altitude was reached the carbon dioxide concentration in exhaust gas was approximately 8%. A reverse relationship can be observed for carbon monoxide concentration. Its initial value amounted to approximately 8% and grew in line with the growth in altitude, up to approximately 10%. In the case of hydrocarbons, their concentration in exhaust gas was constant and amounted to approximately 200 ppm. For nitrogen oxides, the original concentration was approximately 200 ppm, and then went down to approximately 140 ppm as the altitude grew. The above changes in carbon dioxide, carbon monoxide and nitrogen oxides concentrations are related to altitude-induced changes taking place in the combustion process. Changes in ambient pressure caused by changes in altitude reduce the cylinder charging efficiency and – given a constant fuel dose provided (injected) by the fuel supply system – change the excess air coefficient from 0.8 to 0.75. In its turn, this change is equivalent to enriching the fuel/air mixture and lower availability of oxygen in the combustion process and reduces the combustion temperature, which eventually leads to lower emission of nitrogen oxides and greater emission of carbon monoxide. Greater emission of carbon monoxide entails lower emission of carbon dioxide, being a product of full and complete combustion.

After the desired altitude was achieved, the flight stabilized. The procedure of fuel and air mixture adjustment was carried out. Its initial part consisted of maximum leaning of the mixture with simultaneous observation of engine head temperature changes, followed by slight enrichment of the mixture. In this way, with stabilized flight and engine parameters, the excess air coefficient of the fuel and air mixture reached 1.1. At this excess air coefficient, the fuel and air mixture combustion process is characterized by maximized thermal efficiency, which is reflected in the fuel consumption rate.

In a particular manner it is evidence by changes in nitrogen oxides and carbon monoxide emission rates, which in their turn cause an increase in carbon dioxide emission rate. To assess the degree of fuel saving the emission rates of carbon compounds in exhaust gas were analyzed. The analysis consisted in determining the fuel consumption coefficient (FC_c) by means of aggregating the instantaneous emission rates of each compound with the corresponding coefficient depending on the number of carbon atoms in the binding, as shown in the following equation:

$$FC_c = 0.273 \cdot E_{CO_2} + 0.429 \cdot E_{CO} + 0.866 \cdot E_{HC} \text{ [g/s]} \tag{2.13}$$

The results of the analysis of carbon compounds in exhaust gas show that despite higher carbon dioxide emissions resulting from changes in the combustion process caused by an adjustment of the fuel and air mixture, the fuel consumption coefficient FC_c was reduced by approximately 20% (Figure 2.44). The reduction of the fuel consumption coefficient is consequently related to the decreased instantaneous fuel consumption, which confirms the importance of in-flight fuel and air ratio adjustment.

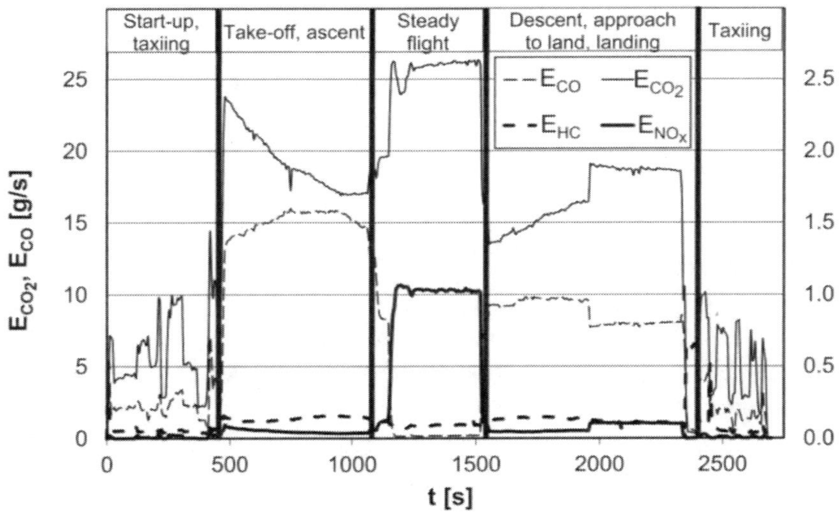

Figure 2.43. Toxic compounds emission rate in each flight phase.

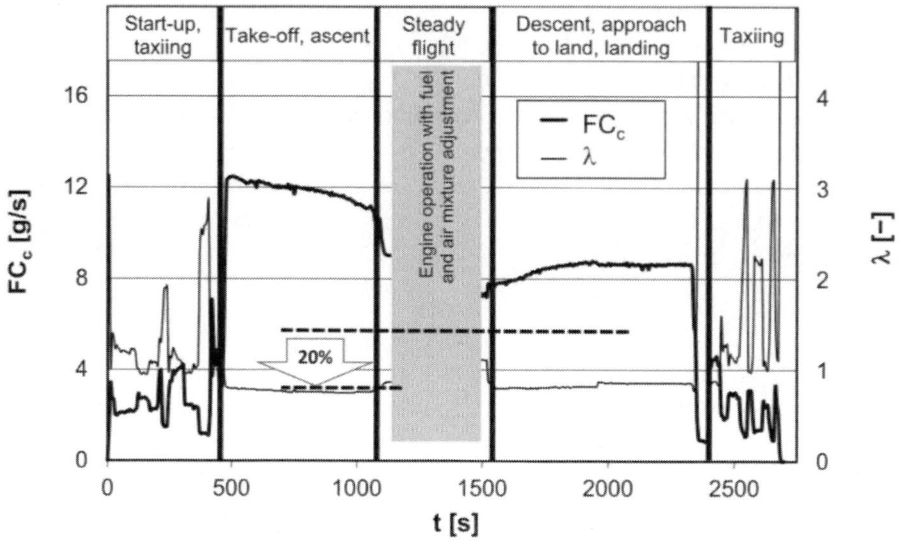

Figure 2.44. Fuel consumption coefficient changes in each flight phase.

In the initial steady flight phase with the excess air coefficient $\lambda = 0.87$, concentrations of carbon dioxide, carbon monoxide, nitrogen oxides and hydrocarbons amounted to approximately 10%, 7%, 500 ppm and 280 ppm, respectively. Following the fuel and air mixture adjustment from rich (with excess air coefficient $\lambda = 0.87$) to lean, nearly stoichiometric (excess air coefficient $\lambda = 1.1$), carbon dioxide concentration grew to approximately 13%, carbon monoxide concentration fell to 0.15%, nitrogen oxides concentration grew to approximately 4500 ppm, and hydrocarbons concentration dropped to approximately 220 ppm.

Modification of the fuel and air mixture composition resulted in a reduction of carbon monoxide concentration by approximately 40-fold, which translates into 1.3-fold increase in carbon dioxide concentration, 10-fold increase in nitrogen oxides concentration and unchanged concentration of hydrocarbons. Changes in emission rates of each compound follow a similar pattern.

The parameters of the fuel supply system used in the steady flight phase are dedicated to airplane operation conditions at a specific altitude and the corresponding static pressure. Unfortunately, changes in flight altitude result in modification of the combustion process in the engine's cylinder, which means that the fuel and air mixture adjustment procedure should be repeated. A situation like this was taken into account during the test. Namely, descent was one of the elements of the test flight. Once the airplane began to descend, the engine load was reduced and the fuel supply system was switched back to the rich air and fuel mixture position ($\lambda = 0.8$), which resulted in changes in the concentration of toxic compounds in exhaust gas. The concentrations of both carbon dioxide and carbon monoxide amounted to approximately 9%, while the approximate concentrations of nitrogen oxides and hydrocarbons were 210 ppm and 370 ppm, respectively. These values were maintained during the descent, and when the flight parameters stabilized again at an altitude of approximately 500 m, the engine load was increased, which caused a growth in carbon dioxide concentration to approximately 10%, reduction in carbon monoxide concentration to approximately 6.5%, growth in nitrogen oxides concentration to approximately 450 ppm, while the concentration of hydrocarbons dropped to approximately 270 ppm. A short steady flight at this altitude was part of the approach procedure, followed by the landing phase characterized by major altitude reduction with the engine operating at no load (or even negative load, as in back-throttling). Under such conditions, the fuel supply system was switched to its lowest position. With a closed throttle, the engine speed induced by the reverse operation of the propulsion system caused forced charging of the cylinder with a limited amount of fuel which – due to the high excess air coefficient – was not burnt. As a matter of consequence, the concentration of hydrocarbons in exhaust gas grew to approximately 3850 ppm, but the concentrations of the remaining toxic compounds were significantly reduced.

After the landing and during the airplane's taxiing to the hardstand the engine's parameters are characteristic for low loads. Toxic compound concentrations are highly variable and their pattern is similar to that observed during taxiing for take-off.

All operation and emission parameters recorded during the test were analyzed in order to calculate instantaneous emission values of each toxic compound. Subsequently those values were aggregated in order to determine the mass of compounds emitted during the 45-minute flight.

It was determined that during the entire operation the airplane's propulsion system emitted 18.8 kg of carbon monoxide, 0.4 kg of nitrogen oxides, 41 kg of carbon dioxide and 0.3 kg of hydrocarbons. These values were subsequently compared against pollution masses derived from the above-described model (Table 2.19).

Table 2.19. Pollution mass model inputs and their empirical equivalents measured during the flight of Cirrus SR20

Parameter	Flight values	Model input values
$N_{e\ max}$ [kW]	145	145
t_{fly} [s]	2689	2689
h_{fly} [m]	1600	1600
v_{asc} [m/s]	4.2	4.2
v_{dsc} [m/s]	3	3
e_{CO} [g/kWh]	–	434
e_{HC} [g/kWh]	–	5.06
e_{NOx} [g/kWh]	–	2.19
e_{CO2} [g/kWh]	–	775
t_{op} [s]	1409	1409
Masses of toxic compounds – in-flight values and model values		
m_{CO} [g]	18 847	15 679
m_{HC} [g]	311	183
m_{NOx} [g]	402	79
m_{CO2} [g]	41 041	27 250

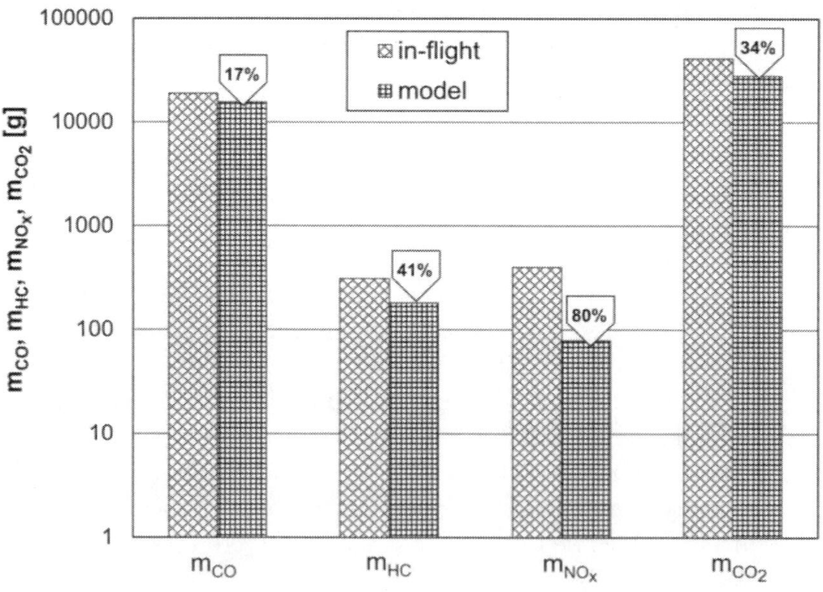

Figure 2.45. Comparison of modeled masses and masses measured in-flight.

The comparison of model emission mass values against in-flight mass emission values is presented visually in Figure 2.45. The figure shows that the relative difference between the model value and the actual value for each compound is significant and results from the pre-flight test methodology and the specific nature of the fuel supply system parameters (no fuel and air mixture adjustment possible). For carbon monoxide mass the relative difference amounts to 17%, as compared to 41% for hydrocarbons, 80% of nitrogen oxides and 34% for carbon dioxide. The greatest difference is observed in the mass of nitrogen oxides, which reflects the adjustment of the fuel and air mixture in actual (in-flight) conditions.

An analysis of concentration and emission rate changes during the flight shows that hydrocarbon emissions are the least susceptible to changes in the fuel and air mixture composition. On that basis one can conclude that the overall percentage difference between the proposed models on the one hand and actual emission on the other is approximately 40%. A comparison of computational values derived from the proposed mass models for each compound against actual values recorded during the flight points out to the need for fine-tuning of the models by means of corrective coefficients.

2.8.3. Correction of Aviation Emission Models

In order to improve the proposed emission mass models an analysis of mutual relationships between various model components was carried out. The model's components correspond to operational conditions of the propulsion system during the pre-flight test and during the airplane's actual operation (before, during and after the flight). To capture those operational changes that are caused by fuel and air mixture adjustment (which leads to significant changes in the emission rate of toxic compounds), the part of the model that addresses the steady flight phase was further developed by introducing the duration of flight with fuel and air mixture adjustment (t_{λ}) as well as the emission change coefficient for each compound (w_{λ}) (resulting from changes in the excess air coefficient). Furthermore, correction coefficient (k_T) applicable to the entire model was also introduced to address the specific methodology of the pre-flight test and the specific emissions of each compound obtained on its basis. The improved model formulae are presented below, and coefficient values are presented in Table 2.20.

Corrected aviation piston engine emission models:

- mass of carbon monoxide [g]:

$$m_{\mathrm{CO}} = \frac{e_{\mathrm{CO}} \cdot N_{e\max} \cdot k_{T\,\mathrm{CO}}}{3600} \cdot R \tag{2.14}$$

- mass of hydrocarbons [g]:

$$m_{\mathrm{HC}} = \frac{e_{\mathrm{HC}} \cdot N_{e\max} \cdot k_{T\,\mathrm{HC}}}{3600} \cdot R \tag{2.15}$$

- mass of nitrogen oxides [g]:

$$m_{\mathrm{NOx}} = \frac{e_{\mathrm{NOx}} \cdot N_{e\max} \cdot k_{T\,\mathrm{NOx}}}{3600} \cdot R \tag{2.16}$$

- mass of carbon dioxide [g]:

$$m_{\mathrm{CO_2}} = \frac{e_{\mathrm{CO_2}} \cdot N_{e\max} \cdot k_{T\,\mathrm{CO_2}}}{3600} \cdot R \tag{2.17}$$

where:

$$R = \frac{h_{fly}}{v_{asc}} + 0.7 \cdot \left(t_{fly} - h_{fly} \cdot \left(\frac{1}{v_{asc}} + \frac{1}{v_{dsc}} \right) + t_\lambda \;\; w_\lambda - 1 \right) + 0.2 \cdot \left(\frac{h_{fly}}{v_{dsc}} + t_{op} \right)$$

Table 2.20. Emission change coefficient w_λ and model correction coefficient k_T for each exhaust gas compound

Gas compound	k_T	w_λ
CO	0.31	0.025
HC	0.6	1
NO_x	1.05	10
CO_2	0.437	1.3

The improved emission mass models were once again verified by comparing them against in-flight values recorded during the flight of Cirrus SR20 (Table 2.21).

The comparison of mass values derived from the improved models against in-flight mass emission values is presented visually in Figure 2.46. It shows that after correcting the model, values calculated on its basis largely correspond to actual values for each compound recorded during the airplane's operation. This allows one to conclude that the new coefficients were properly introduced into the model and their respective values are correct.

Table 2.21. Input values for corrected emission mass models and in-flight values recorded during the flight of Cirrus SR20

Parameter	Flight values	Model input values
$N_{e\,max}$ [kW]	145	145
t_{fly} [s]	2689	2689
h_{fly} [m]	1600	1600
v_{asc} [m/s]	4.2	4.2
v_{dsc} [m/s]	3	3
e_{CO} [g/kWh]	–	434
e_{HC} [g/kWh]	–	5.06
e_{NOx} [g/kWh]	–	2.19
e_{CO2} [g/kWh]	–	775
t_{op} [s]	1409	1409
t_{λ} [s]	372	372
$w_{\lambda\,CO}$	–	0.025
$w_{\lambda\,HC}$	–	1
$w_{\lambda\,NOx}$	–	10
$w_{\lambda\,CO2}$	–	1.3
$k_{T\,CO}$	–	0.31
$k_{T\,HC}$	–	0.6
$k_{T\,NOx}$	–	1.05
$k_{T\,CO2}$	–	0.437
Masses of toxic compounds – in-flight values and model values		
m_{CO} [g]	18 847	18 875
m_{HC} [g]	311	312
m_{NOx} [g]	402	401
m_{CO2} [g]	41 041	41 057

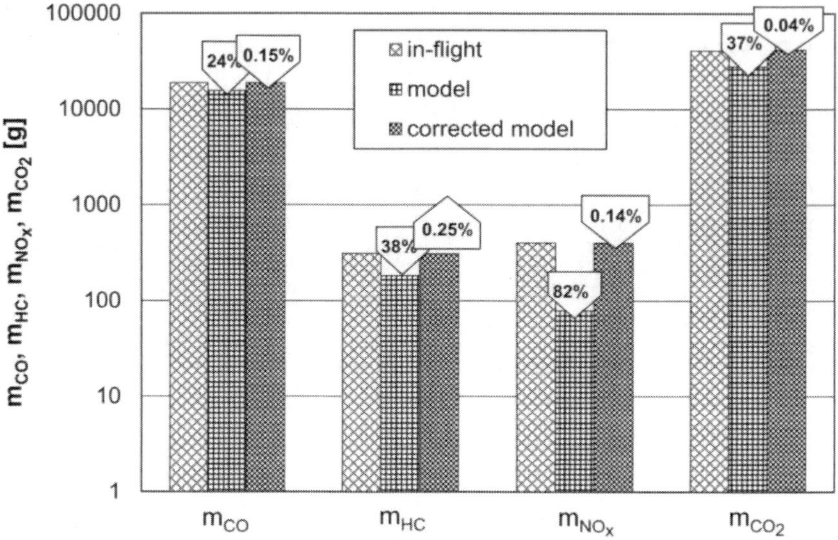

Figure 2.46. Comparison of modeled masses (before and after correction) and masses measured in in-flight conditions.

The corrected models were once again verified in a comparative analysis involving the masses of toxic compounds recorded during another flight. The following subchapter of this dissertation discusses the details of that analysis.

Modifications of the original emission mass models require an additional verification of results derived from the models after such modifications. The verification was carried out on the basis of toxic compound emission values recorded in in-flight conditions. The comparative analysis featured data from emission tests carried out during another flight of the Cirrus SR20, taking place on the same day as the flight used in the previous analysis. The second flight began one hour after the first one. The airplane was operated by the same pilot, and the same measurement instrumentation was used. Flight duration and altitude were intentionally different from those in the first flight. It was assumed that the fuel and air mixture composition would not be adjusted during the flight. Since the flight altitude was supposed to be approximately 500 m, the airplane flew in the same direction as before and turned back above the town of Oborniki, in order to approach the Poznań-Ławica airport from the west. The entire operation began and ended in the airplane's hangar. The flight was completed as planned, measurements of toxic emissions in exhaust gas were successfully recorded, and so were the operational parameters of the airplane and its propulsion system. Among the necessary environmental and operational performance parameters, the following values were monitored: concentrations of toxic compounds in exhaust gas (carbon monoxide, hydrocarbons, nitrogen oxides and carbon dioxide), exhaust gas flow rate, engine speed, flight altitude, as well as geographical latitude and longitude. The recorded GPS signal served as a basis for visualizing the airplane's trajectory (Figure 2.47).

The engine's operating conditions are closely related to the nature of changes in flight altitude and fuel supply system parameters (Figure 2.48) and are reflected by the concentration of toxic pollutants in exhaust gas (Figure 2.49). The function plots presented below reflect the nature of changes in the concentration of each pollutant, connected with operating conditions resulting from changes in flight altitude.

The test's results denote a close correlation of the engine's performance parameters with changes in the concentrations and emission rates of toxic compounds in exhaust gases. For start-up, warming up, taxiing for take-off and hardstand, the engine operates at highly variable speeds and low loads, which is reflected in changing toxic emission rates. In the other flight phases, the engine and airplane performance values are fairly constant. This means that the take-off ascent had a constant speed at maximum fuel supply to the engine and at maximum engine power. Concentrations of toxic compounds in exhaust gas in the take-off and ascent phases exhibit the following values: from the beginning of the take-off the carbon dioxide concentration amounted to approximately 10% and then gradually decreased as the altitude grew. After the desired altitude (approx. 500 m) was reached the carbon dioxide concentration in exhaust gas was approximately 8%. For carbon monoxide concentration, a reverse relationship can be observed. Its initial value amounted to approximately 7% and grew in line with the growth in altitude, up to approximately 8%. For hydrocarbons, their concentration in exhaust gas was constant and amounted to approximately 150 ppm. As regards nitrogen oxides, the initial concentration at the beginning of take-off was approximately 600 ppm, and then plunged to approximately 250 ppm. Subsequently it was further reduced in as the altitude grew, to approximately 200 ppm. The above changes in carbon dioxide, carbon monoxide and nitrogen oxides concentrations are related to altitude-induced changes taking place in the combustion process. After the desired altitude was achieved, the flight stabilized and the concentrations of toxic compounds took the following values: approximately 6% for carbon monoxide, approximately 150 ppm for hydrocarbons, approximately 450 ppm for nitrogen oxides and approximately 9% for carbon dioxide. During landing the engine load was reduced, which resulted in lower exhaust gas concentrations of carbon monoxide, nitrogen oxides and carbon dioxide, as well as an increase in the concentration of hydrocarbons (up to approximately 4000 ppm). After the airplane landed and began to taxi, toxic compound concentrations became highly variable, not unlike during taxiing for take-off. In this phase, the carbon monoxide concentration range is 2–6%, as compared to 150–1000 ppm for hydrocarbons, 20–300 ppm for nitrogen oxides, and 3–9% for carbon dioxide. All operation and emission parameters recorded during the test were analyzed, which made it possible to calculate and aggregate instantaneous emission values of each toxic compound. The total mass of emitted compounds was computed on the basis of the entire flight duration (Figure 2.50). In this way it was determined that during the entire flight (including airport operations) the airplane emitted 12.4 kg of carbon oxide, 0.2 kg of hydrocarbons, 0.1 kg of nitrogen oxides and 27 kg of carbon dioxide. These values were subsequently compared against pollution masses derived from the emission mass model (Table 2.22).

Figure 2.47. Airplane trajectory during the test flight; taxi to take-off (blue), take-off and ascent (red), steady flight (purple), approach to land and landing (yellow), taxi to hangar (light blue).

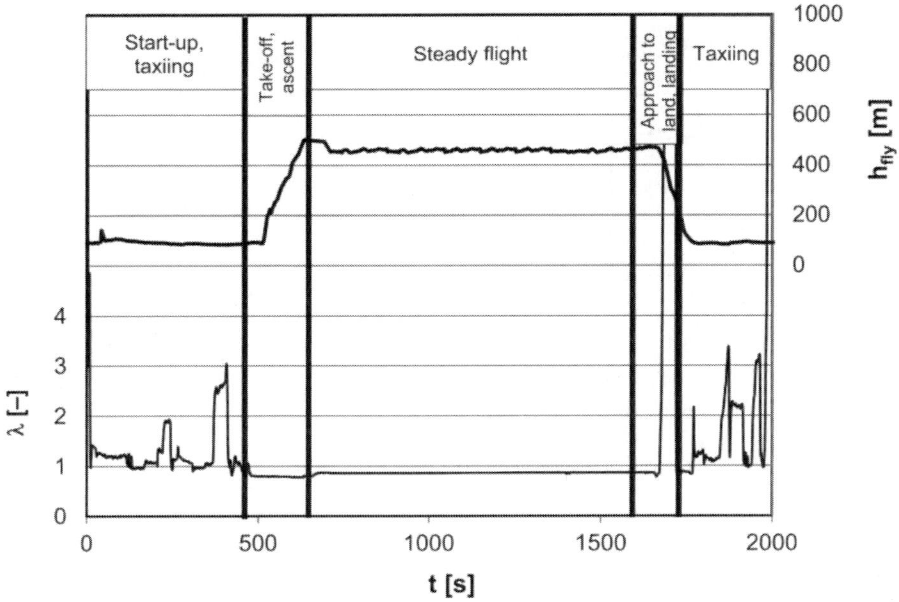

Figure 2.48. Changes in flight altitude and excess air coefficient during the flight.

Table 2.22. Input values for emission mass models and in-flight values recorded during the flight of Cirrus SR20

Parameter	Flight values	Model input values
$N_{e\,max}$ [kW]	145	145
t_{fly} [s]	1990	1990
h_{fly} [m]	500	500
v_{asc} [m/s]	4.2	4.2
v_{dsc} [m/s]	3	3.0
e_{CO} [g/kWh]	–	434
e_{HC} [g/kWh]	–	5.06
e_{NOx} [g/kWh]	–	2.19
e_{CO2} [g/kWh]	–	775
t_{op} [s]	904	904
Masses of toxic compounds – in-flight values and model values		
m_{CO} [g]	12 392	12 289
m_{HC} [g]	217	214
m_{NOx} [g]	138	140
m_{CO2} [g]	27 298	26 674

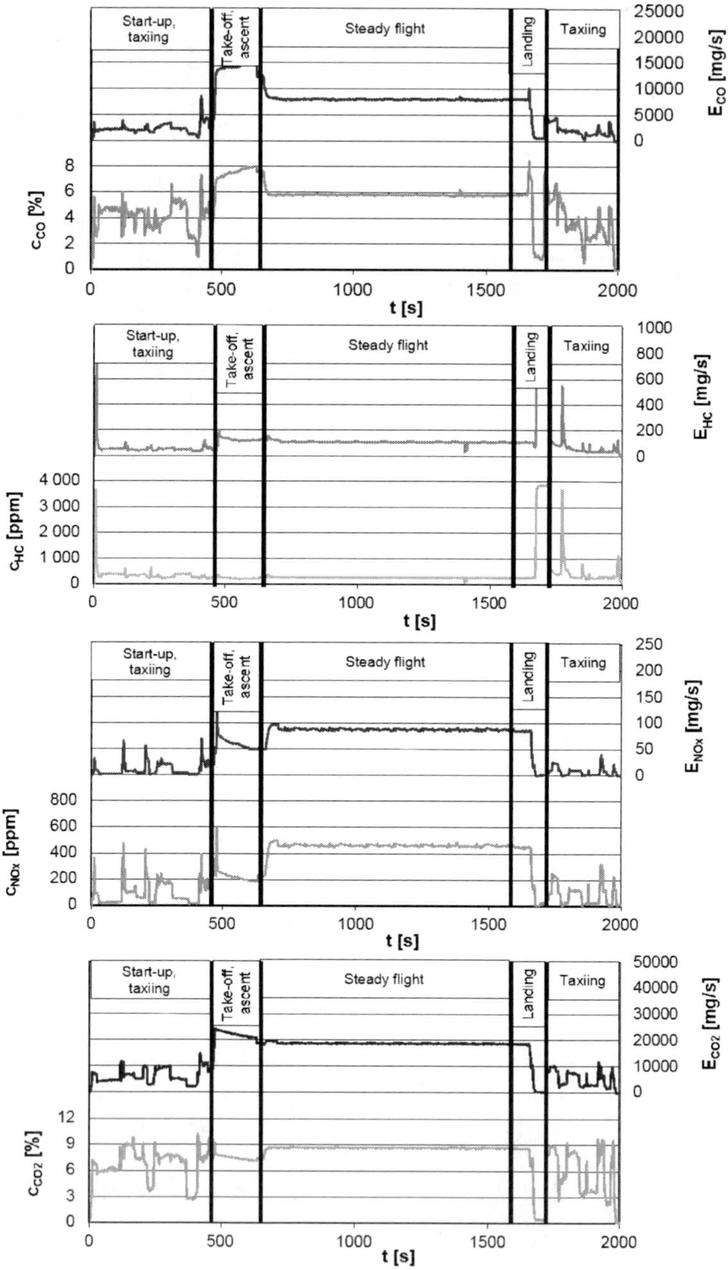

Figure 2.49. Changes in toxic compound concentrations and emission rates in exhaust gas in each flight phase.

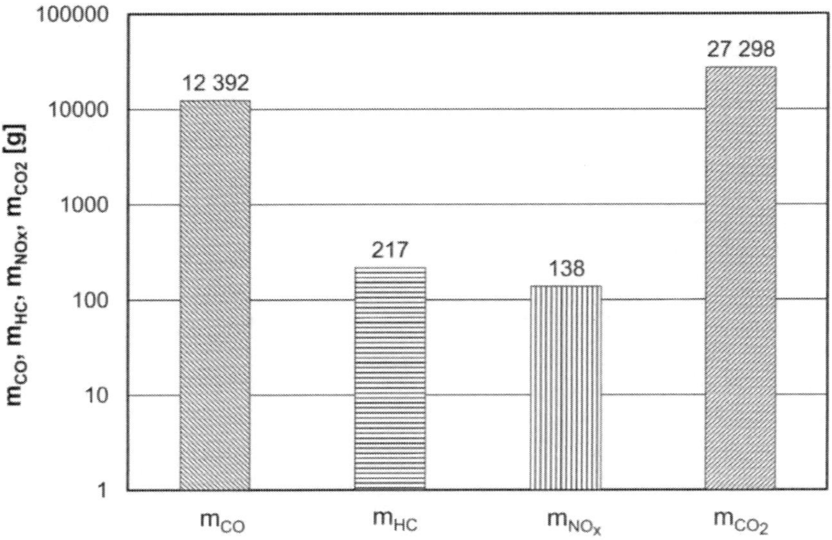

Figure 2.50. Mass of toxic compounds in exhaust gas emitted during the flight.

The comparison of mass values derived from the improved models against in-flight mass emission values is presented in Figure 2.51, showing that the relative difference between the model value and the actual value for each compound is insignificant. For carbon monoxide mass the relative difference amounts to 0.83%, as compared to 1.54% for hydrocarbons, 1.36% of nitrogen oxides and 2.29% for carbon dioxide. These results indicate that the model structure is appropriate and suits a variety of operating conditions of piston engines. The greatest difference is observed in the mass of carbon dioxide and nitrogen oxides, which may result from certain differences between actual (in-flight) conditions and those assumed for the model (e.g., gusts of wind, changing atmospheric pressure or wind directions, etc.). However, none of the differences is greater than 5%, which indicates that the developed emission mass models are highly accurate.

As a result, one is justified to conclude that the above emission mass models make it possible to determine and assess the amount of toxic compounds in exhaust gas generated by aviation piston engines in in-flight conditions. The said emission mass models were used for such assessment with regard to the propulsion systems of all aircraft participating in pre-flight tests. Selected results of those assessments are analyzed in the following chapter of this work.

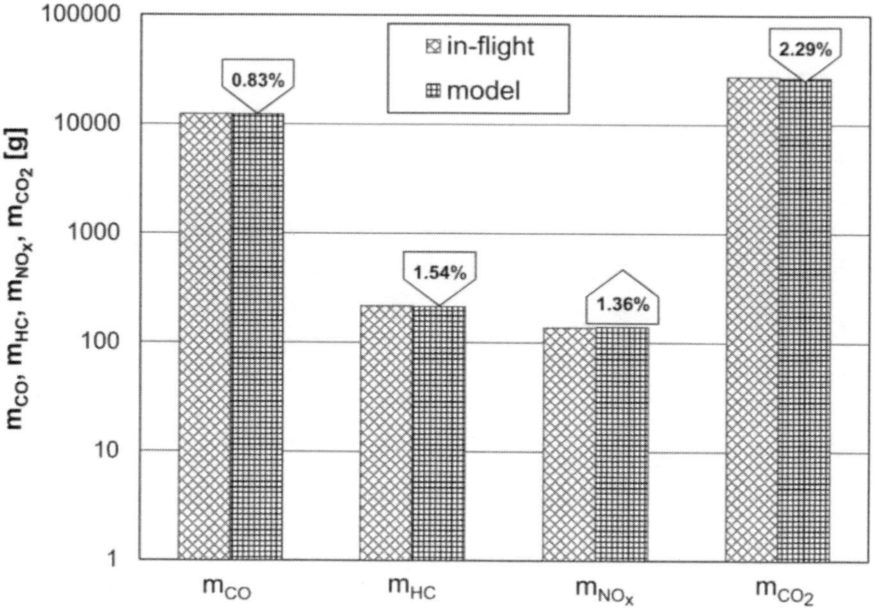

Figure 2.51. Comparison of modeled masses and masses measured in-flight.

The models were used to determine the mass of toxic compounds generated by the propulsion systems of aircraft participating in the proposed stationary emission test (Table 2.23). Similarly to the previous analysis, the same flight parameters were assumed for all airplanes, namely: flight altitude h_{fly} = 500 m, flight duration t_{fly} = 3600 s), and unified time of airport operations consisting of engine start-up, warming up, cooling down and taxiing (t_{op} = 800 s). The maximum effective engine power $N_{e\,max}$ was assumed on the basis of each engine's characteristics. Ascent velocity during take-off and descent velocity during landing were assumed on the basis of technical specifications of the airplanes. For all analyzed cases it was assumed that the fuel and air mixture composition was not adjusted during the steady flight phase (t_{λ} = 0). The values of specific emissions of individual compounds, considered as model input parameters, were determined on the basis of emission measurements carried out under the proposed test conditions. Emission mass models enhanced with corrective coefficients were applied. The results are presented visually in Figure 2.52.

The next step consisted of calculating the mass of toxic compounds that would be emitted by an airplane's propulsion system if the fuel and air mixture parameters were adjusted during the steady flight phase (assumed steady flight duration t_λ = 2000 s). The model's input values are presented in Table 2.24, and the resultant values are visualized in Figure 2.53.

By comparing emission mass values for each compound, one can conclude that the model adequately reflects changes triggered by fuel and air mixture adjustment. These changes include reduced mass of carbon monoxide and increased mass of nitrogen oxides, accompanied by slightly greater mass of carbon dioxide and unchanged mass of hydrocarbons (Figure 2.54).

Table 2.23. Input values for emission mass models and resultant values for selected airplanes

Parameter	Airplane				
	PZL-104 Wilga 35	**Cessna 152 II**	**Zlin-142 M**	**Cirrus SR22T (SP-IES)**	**Cirrus SR20**
Input values					
$N_{e\,max}$ [kW]	194	86	185	233	145
t_{fly} [s]	3600				
h_{fly} [m]	500				
v_{asc} [m/s]	5.0	3.6	4.4	6.0	4.2
v_{dsc} [m/s]	3.0	3.0	2.7	3.0	3.0
e_{CO} [g/kWh]	239	327	101	141	434
e_{HC} [g/kWh]	12.3	8.97	1.47	4.10	5.06
e_{NOx} [g/kWh]	2.05	2.32	2.02	0.31	2.19
e_{CO2} [g/kWh]	585	766	540	524	775
t_{op} = 800 s, t_λ = 0 s $w_{\lambda\,CO}$ = 0.025, $w_{\lambda\,HC}$ = 1, $w_{\lambda\,NOx}$ = 10, $w_{\lambda\,CO2}$ = 1.3 $k_{T\,CO}$ = 0.31, $k_{T\,HC}$ = 0.6, $k_{T\,NOx}$ = 1.05, $k_{T\,CO2}$ = 0.437					
Pollutant mass values derived from the models					
m_{CO} [g]	13 094	8169	5356	9159	18 033
m_{HC} [g]	1122	368	128	447	347
m_{NOx} [g]	303	153	286	55	242
m_{CO2} [g]	41 392	24 520	36 644	44 137	41 398

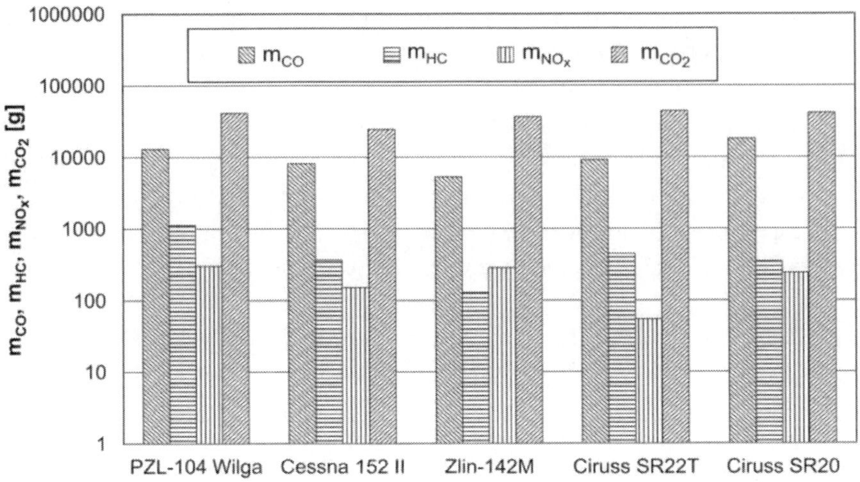

Figure 2.52. Comparison of model-based emission masses for assumed operation parameters (no fuel and air mixture composition adjustment).

Table 2.24. Input values for emission mass models and resultant values for selected airplanes (with fuel and air mixture composition adjustment)

Parameter	PZL-104 Wilga 35	Cessna 152 II	Zlin-142 M	Cirrus SR22T	Cirrus SR20
Input values					
$N_{e\,max}$ [kW]	194	86	185	233	145
t_{fly} [s]	3600				
h_{fly} [m]	500				
v_{asc} [m/s]	5.0	3.6	4.4	6.0	4.2
v_{dsc} [m/s]	3.0	3.0	2.7	3.0	3.0
e_{CO} [g/kWh]	239	327	101	141	434
e_{HC} [g/kWh]	12.3	8.97	1.47	4.10	5.06
e_{NOx} [g/kWh]	2.05	2.32	2.02	0.31	2.19
e_{CO2} [g/kWh]	585	766	540	524	775
$t_{op} = 800$ s, $t_\lambda = 2000$ s $w_{\lambda\,CO} = 0.025$, $w_{\lambda\,HC} = 1$, $w_{\lambda\,NOx} = 10$, $w_{\lambda\,CO2} = 1.3$ $k_{T\,CO} = 0.31$, $k_{T\,HC} = 0.6$, $k_{T\,NOx} = 1.05$, $k_{T\,CO2} = 0.437$					
Pollutant mass values derived from the models					
m_{CO} [g]	7644	4868	3149	5297	10 636
m_{HC} [g]	1122	368	128	447	347
m_{NOx} [g]	1765	886	1668	320	1409
m_{CO2} [g]	47 178	27 879	41 737	50 362	47 128

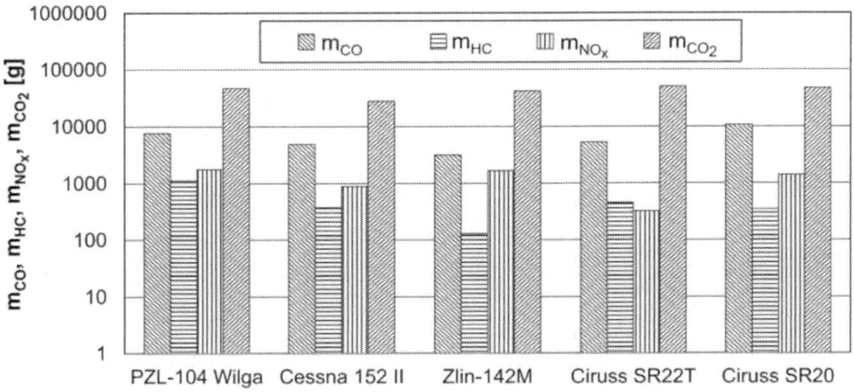

Figure 2.53. Comparison of model-based emission masses for assumed operation parameters (with fuel and air mixture composition adjustment).

For each toxic compound the above changes can be described by means of percentage values, which are as follows (Figure 2.55): carbon monoxide mass fell by approximately 41%, hydrocarbons mass remains unchanged, nitrogen oxides mass grows by approximately 480%, while the mass of carbon dioxide grows by 14%. Changes expressed by means of percentage values may blur the view of the scale of actual emission changes, which is why attention must be drawn to absolute masses of emitted pollutants. After the fuel and air mixture composition adjustment in the Zlin-142 M, the mass of emitted carbon monoxide falls from 5356 g to 3149 g; nitrogen oxides increase from 286 g to 1668 g, and carbon dioxide goes up from 36 644 g to 41 737 g.

From the environmental perspective, reduction in carbon monoxide mass is the most important. During a 1-hour flight of the Zlin-142 M with fuel and air mixture composition adjustment, CO emission can be reduced by approximately 3000 g. The growth in nitrogen oxides (by approx. 1380 g) is also important. For aircraft, the environmental impact of nitrogen oxides depends on the flight altitude, as the compound reacts directly with ozone from upper atmospheric layers. In its turn, increased carbon dioxide mass is indicative of improved efficiency of thermal conversion in the engine, which denotes better use of fuel fed into the engine and translates into its lower consumption.

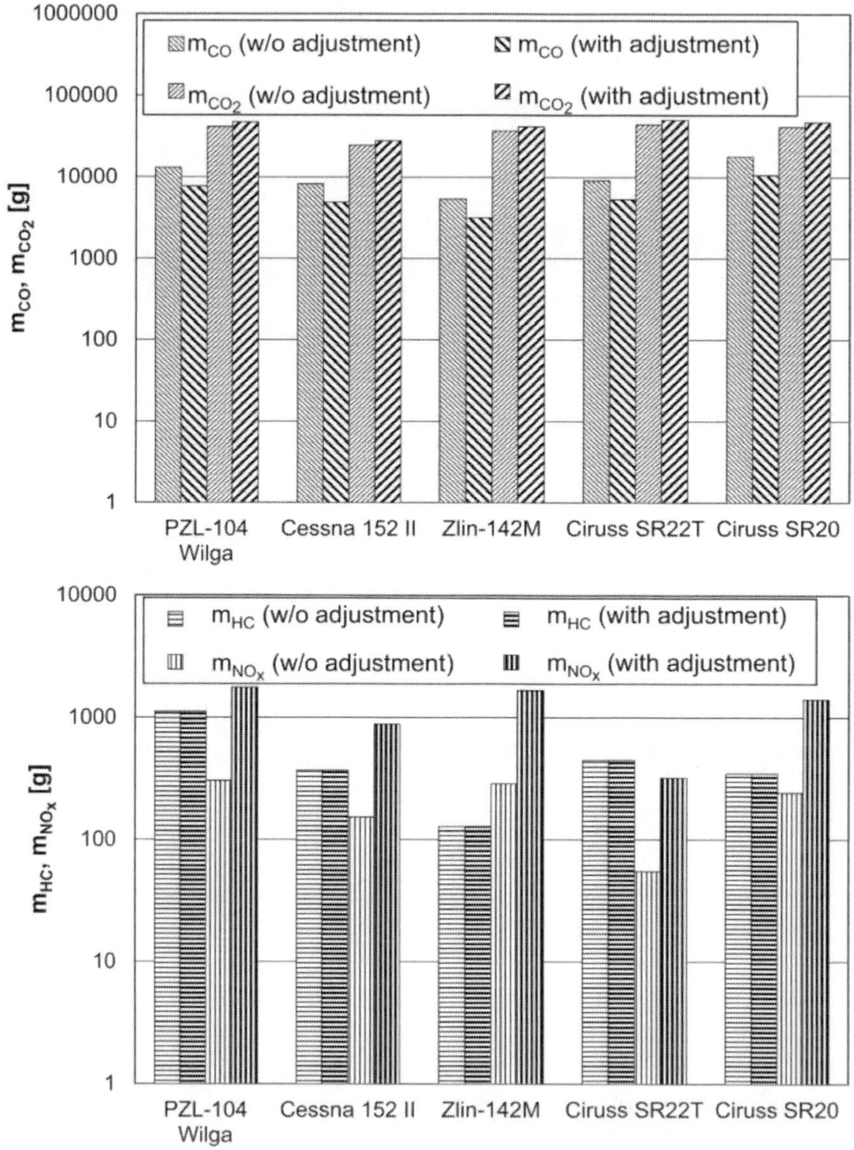

Figure 2.54. Comprehensive comparison of model-based emission masses for assumed airplane operation parameters (with and without fuel and air mixture composition adjustment).

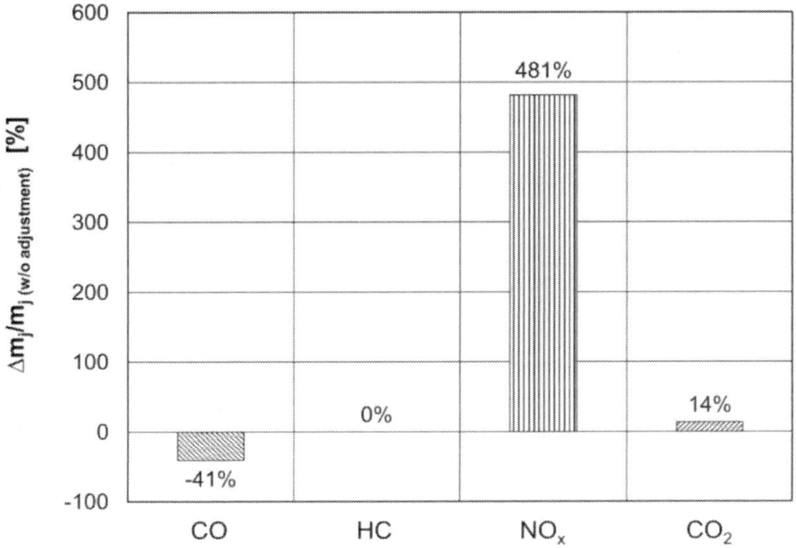

Figure 2.55. Change in model-derived emission values for assumed airplane operation parameters resulting from fuel and air mixture composition adjustment.

Table 2.25. Respective durations of all circuit pattern legs for Zlin-142 M, making up the times of phases featured in the developed emission mass models

Phase 1 of circuit pattern	Take-off	Ascent + 1st turn	Ascent + 2nd turn	Level flight (base leg) + 3rd turn	Descent + 4th turn	Final leg	Soaring, floating, landing braking and stop
Leg time slot [s]	0 – 41	42 – 49	50 – 81	82 – 221	222 – 250	251 – 259	260 – 283
Leg duration[s]	41	8	41	140	29	9	23
Corresponding model flight phase	t_{asc}			t_{steady}	t_{dsc}		
Model phase duration [s]	90			140	61		
Flight duration t_{fly} [s]	291						

Among all types of general aviation operations, training flights are particularly harmful to the environment. Training flights are very frequent and take place at low altitudes. Circuit patterns are an example of such a flight. In accordance with training procedures, circuit patterns consist of several phases (legs), as shown in Tables 2.25 and 2.26. Depending on the airplane's capacity, relative durations of those legs may vary. Knowing the airplane parameters, the most important parameters required for developing emission mass models were identified, including ascending and descending velocity as well as flight duration and altitude.

Table 2.26. Emission mass model input values and resultant emission values for Zlin-142 M flying circuit patterns

Parameter	Zlin-142 M
$N_{e\,max}$ [kW]	185
t_{fly} [s]	291
h_{fly} [m]	400
v_{asc} [m/s]	4.4
v_{dsc} [m/s]	2.7
e_{CO} [g/kWh]	101
e_{HC} [g/kWh]	1.47
e_{NOx} [g/kWh]	2.03
e_{CO2} [g/kWh]	540
t_{op} [s]	0
t_{λ} [s]	0
$w_{\lambda\,CO}$	0.025
$w_{\lambda\,HC}$	1
$w_{\lambda\,NOx}$	10
$w_{\lambda\,CO2}$	1.3
$k_{T\,CO}$	0.31
$k_{T\,HC}$	0.6
$k_{T\,NOx}$	1.05
$k_{T\,CO2}$	0.437
Pollutant mass values derived from the models	
m_{CO} [g]	687
m_{HC} [g]	11
m_{NOx} [g]	17
m_{CO2} [g]	3786

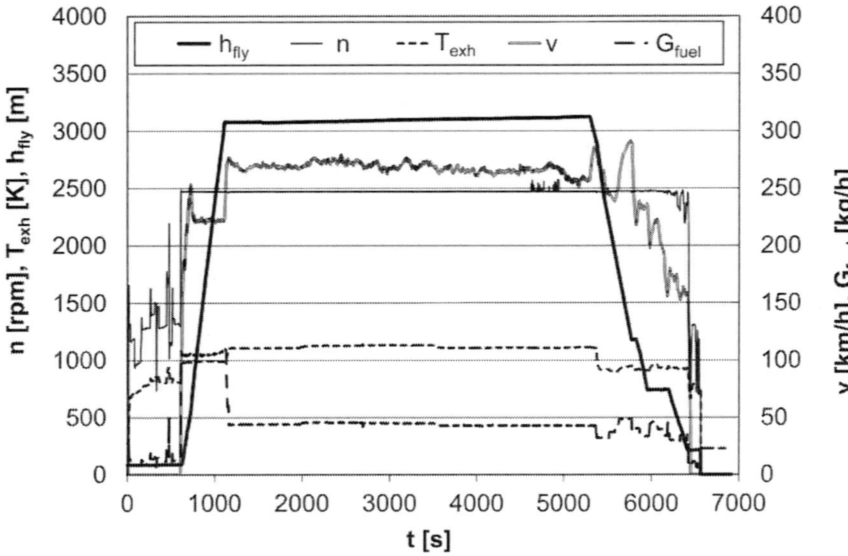

Figure 2.56. Selected operational parameters of Cirrus SR22T and its propulsion system during the flight from Poznań-Ławica to Rzeszów-Jasionka.

By comparing the above data and assuming that the flight in question is a circuit pattern excluding airport operations (t_{op} = 0 s) and that it is not possible to adjust the fuel and air mixture composition (t_{λ} = 0 s), the mass of toxic compounds was determined on the basis of the proposed models (Table 2.26). A circuit pattern analysis carried out for the Zlin-142 M gave the following emission results: 687 g of carbon dioxide, 11 g of hydrocarbons, 17 g of nitrogen oxides and 3786 g of carbon dioxide. Assuming than a single trainee flies 10 circuit patterns and that 10 different trainees practice on the same day one can calculate that in the vicinity of the airport and at a relatively low altitude the propulsion system of the Zlin-142 M will generate 68.7 kg of carbon monoxide, 1.1 kg of hydrocarbons, 1.7 kg of nitrogen oxides and 378.6 kg of carbon dioxide. The mass of toxic compounds calculated for a training airplane flown repeatedly in the vicinity of an airport allows one to realize the hazards for the natural environment and for local residents.

The above analytical methodology can also be used for assessing actual emissions of toxic compounds from general aviation aircraft during commercial flights. A good example here is the flight of Cirrus SR22T from the Poznań-Ławica airport to the destination airport of Rzeszów-Jasionka. During the flight, airplane and propulsion system operational parameters were recorded, based on which it is possible to assess the flight's character and determine the duration of each flight phase (Figure 2.56).

On the basis of the typical flight analysis presented in chapter 3.2 one can determine that the total duration of the flight in question including taxiing after landing and engine cooling is t_{fly} = 6565 s, or approximately 110 minutes. The airplane took the shortest possible route and flew at the intended altitude h_{fly} = 3080 m, with fixed parameters of the engine fuel feed. The identified phases also include start-up, warming up, taxi for take-off and hardstand, as well as engine cooling – collectively referred to as ground operations – whose total duration was t_{op} = 716 s. The steady flight duration was t_{λ} = 4180 s.

Table 2.27. Emission mass model input values and calculated emission masses for Cirrus SR22T

Parameter	Cirrus SR22T
$N_{e\ max}$ [kW]	233
t_{fly} [s]	6570
h_{fly} [m]	3080
v_{asc} [m/s]	6.0
v_{dsc} [m/s]	3.0
e_{CO} [g/kWh]	140
e_{HC} [g/kWh]	4.10
e_{NOx} [g/kWh]	0.31
e_{CO2} [g/kWh]	524
t_{op} [s]	716
t_{λ} [s]	4180
$w_{\lambda\ CO}$	0.025
$w_{\lambda\ HC}$	1
$w_{\lambda\ NOx}$	10
$w_{\lambda\ CO2}$	1.3
$k_{T\ CO}$	0.31
$k_{T\ HC}$	0.6
$k_{T\ NOx}$	1.05
$k_{T\ CO2}$	0.437
Pollutant mass values derived from the models	
m_{CO} [g]	9686
m_{HC} [g]	789
m_{NOx} [g]	646
m_{CO2} [g]	94 423

The airplane in question is fitted with automatic fuel and air mixture adjustment system. After the plane reaches the desired altitude the system – based on the thrust lever position, ambient pressure and exhaust gas temperature upstream of the turbine – adjusts the fuel dose fed into the engine. Adjustment is then confirmed by increased exhaust gas temperature upstream of the turbine during the steady flight phase.

Parameters and specific emission values recorded during the pre-flight test and used as inputs for emission mass models make it possible to calculate the mass of toxic compounds generated during the analyzed flight (Table 2.27).

The analysis of the Cirrus SR22T flight from Poznań-Ławica to Rzeszów-Jasionka shows the masses of compounds emitted during the flight, namely: 9686 g of carbon monoxide, 789 g of hydrocarbons, 646 g of nitrogen oxides and 94 423 g of carbon dioxide.

The above examples of practical usage of emission mass models can help assess the mass of toxic compounds in exhaust gas generated by aviation piston engines and evaluate its actual impact on the environment.

The main goal of this dissertation was to develop a tool for calculating the mass of gaseous pollutants emitted from aviation piston engines. The research question formulated in the dissertation ("How can the emission of hazardous gaseous compounds contained in exhaust gas from aviation piston engines be correlated with operational parameters of such engines in actual conditions?") required a number of operational steps:

- analysis of operational parameters of aviation piston engines,
- development of specific emission tests and their practical methodology for pre-flight tests,
- development of emission mass models covering actual aircraft operation ranges (particularly during the flight),
- verification of emission mass models against in-flight measurements.

Each of the above operational steps contributed to developing and proposing emission mass models dedicated to general aviation piston engines, thus making it possible to identify and assess actual (in-flight) emissions. The models are based on input data from aircraft operation specifications affecting their actual operating conditions, as well as on specific emission values of each toxic compound measured by means of purpose-designed stationary test procedures that can be integrated with pre-flight tests. The proposed models were verified against exhaust gas emissions measured in in-flight conditions. The comparison of model-derived emission masses against masses measured during the flight confirm a high degree of modeling accuracy. The developed emission mass models were then used to determine specific emissions for selected airplanes, previously subjected to the pre-flight test. The above analyses were carried out for a number of operation variants typically used in general aviation.

Studies into operating conditions of general aviation aircraft as well as analyses of operating parameters of aviation piston engines, coupled with identification and assessment of hazardous gaseous emissions from internal combustion engines made it possible to evaluate the range and methods of their use. The following methodological conclusions on testing emissions from aviation piston engines may provide a basis for selecting such analytical directions that are optimally aligned with one's research goals:

1. Operational parameters of aviation piston engines are characterized by little variability during the flight.
2. Aviation piston engines are largely operated at a low excess air coefficient ($\lambda \approx 0.8$), which is determined by the need for internal cooling of the engine by means of low-temperature combustion, which results in elevated emissions of hydrocarbons and carbon monoxide.
3. It is possible to operate aviation piston engines with such parameters of the fuel and air mixture that the excess air coefficient $\lambda \approx 1.1$ (in steady-flight conditions at partial engine load).
4. The procedures of each flight phase are repetitive and depend on the airplane's structural capabilities.
5. In-flight operation conditions of aviation engines depend on propulsion system parameters and can be recreated in pre-flight conditions.
6. It is possible to develop an all-purpose stationary emission test dedicated to aircraft with piston engines and carried out during the pre-flight test, thus enabling determination of emission limits for the purpose of developing standards on emissions from aviation piston engines.

The Authors' analysis of issues in identifying toxic compounds contained in exhaust gas from aviation piston engines serves as a basis for formulating the following substantive conclusions:

1. Efforts aimed at reducing specific carbon monoxide emissions from aviation piston engines are desirable. The current emission levels constitute direct threat for the airplane's pilot and crew.
2. A proprietary method of in-service measurement of emissions from small general aviation aircraft has been developed and may serve as a basis for assessing environmental performance of such aircraft.
3. The proposed emission standards for each compound and the method of their verification in stationary tests may significantly contribute to developing aviation piston engines with improved technical and environmental performance.

In the light of available literature, the direction of studies initiated by the Authors is well aligned with key publications dealing with the issue of toxic emissions from aviation piston engines. The main research directions include qualitative and quantitative assessment of toxic compounds emitted by internal combustion engines. Ideally, they should lead to their reduction. Further research into emissions will be aimed at developing specific emission standards and their presentation to organizations empowered to implement such standards. It will also be targeted at including particulate matter in the scope of specific emission assessment. In this regard, studies aimed at developing models presenting particulate matter emissions from general aviation aircraft in in-flight conditions are anticipated. The next likely step is development of a collective model (based on the proposed emission mass models) for the purpose of determining the mass of pollutions generated by general aviation aircraft.

REFERENCES

[1] Agarwal, R. K. (2009). Sustainable (green) aviation: challenges and opportunities. *SAE Technical Paper Series*, 2009-01-3085.

[2] Brankovic, A., Ryder, R. C., Hendricks, R. C. & Huber, M. L. (2007). A step towards CO_2-neutral aviation. *SAE Technical Paper Series*, 2007-01-3790.

[3] Brasseur, G. P., Cox, R. A., Hauglustaine, D., Isaksen, I., Lelieveld, J., Lister, D. H., Sausen, R., Schumann, U., Wahner, A. & Wiesen P. (1988). European scientific assessment of the atmospheric effects of aircraft emissions. *Atmospheric Environment*, *32*, 13, 2329-2418.

[4] Brook, R. D., Franklin, B., Cascio, W., Hong, Y., Howard, G., Lipsett, M., Luepker, R., Mittleman, M., Samet, J., Smith, S. C. & Tager I. (2004). Air pollution and cardiovascular disease. *Circulation, 109.*

[5] Carruthers, D. (2005). *Cambridge environmental research consultants.* Near Field Dispersion of Aircraft Jet Releases using ADMS. Amsterdam: AERONET Workshop.

[6] Chłopek Z. (2010). Some remarks on engine testing in dynamic states. *Combustion Engines, 143*, 4.

[7] Communication from the Commission to the Council and the European Parliament Thematic Strategy on Air Pollution. Brussels: Commission of the European Communities, 2005.

[8] Dewes, W. et al. (2000). The AERONET Network-Emissions. Atmospheric Impact & Regulations. *Air & Space Europe, 3*, 1-2.

[9] Dockery, D. W., Pope, C. A., Xu, X., Spengler, J. D., Ware, J. H., Fay, M. E., Ferris, B. G. & Speizer, F. E. (2000). An association between air pollution and mortality in six U.S. cities.

[10] Döpelheuer, A. & Lecht, M. (1998). Influence of engine performance on emission characteristics. Lisbon: RTO AVT Symposium on Gas Turbine Engine Combustion, Emissions and Alternative Fuels.

[11] Döpelheuer, A. (2001). Quantities, characteristics and reduction potentials of aircraft engine emissions. *SAE Technical Paper Series*, 2001-01-3008.

[12] Duchene, N. (2005). ALAQS Case Study: Chopin Airport Emission Inventory. Results of the Phase 1 of Contract between Eurocontrol Experimental Centre and Polish Institute of Aviation.

[13] Emission Standards and Test Procedures for Aircraft and Aircraft Engines. *Summary and Analysis of Comments*. EPA420-R-05-004, 2005.

[14] Environmental Protection Agency, Rules and Regulations: §1065.920 *PEMS calibrations and verifications*, Vol. *73*, No. 126, 2008.

[15] Faber, J., Greenwood, D., Lee, D., Mann, M., Mendes de Leon, P., Nelissen, D., Owen, B., Ralph, M., Tilston, J., Velzen, A. & Vreede, G. (2008*). Lower NO_x at higher altitudes policies to reduce the climate impact of aviation NO_x emission.*

[16] Fleuti, E. (2005). *Aircraft ground handling emissions at Zurich airport.* Stockholm: AERONET WorkShop.

[17] Fleuti, E. & Polymeris J. (2004). Unique (Flughafen Zurich AG), *Aircraft NO_x-emissions within the operational LTO cycle.*

[18] Gao, Y. & Checkel M. D. (2007). Emission factors analysis for multiple vehicles using an on-board, in-use emissions measurement system. S*AE Technical Paper Series*, 2007-01-1327.

[19] General Aviation. (2012). *Statistical Databook & Industry Outlook*, GAMA; 2012.

[20] Gonzalez, C. & Jesik, R. L. (2000). The multi-fuel general aviation piston engine. *SAE Technical Paper Series*, 2000-01-1714.

[21] Hausen, R. B., Romano, L. N., Schlosser, J. F., Martins, M. & Trindade, A. Jr. (2010). Efficient ethanol engines in agricultural aviation. *SAE Technical Paper Series*, 2010-36-0398.

[22] Hepburn, C. & Muller, B. (2010). International air travel and greenhouse gas emissions: a proposal for an adaptation levy. *The World Economy*.

[23] Heywood, J. B. (1988). *Internal combustion engine fundamentals*. New York: McGraw Hill.

[24] Janusiewicz, A. (2004). *Integrated assessment of emissions, air quality and health impact related to monitoring results*. Kraków: JRC-Info Day.

[25] Jeż, M. (2007). *Airport environmental impact – methodology of atmosphere protection*. Warsaw: Institute of Aviation Scientific Library.

[26] Kotlarz, W. (2004). *Turbine driving systems as pollution sources at military airports*. Dęblin: Air Forces Academy.

[27] Lelievre, C. P., Ramarouson, R. & Wilson, C.W. (2005). *Volatile organize compounds (VOCs) and the impact on airport air quality. AERONET III.*

[28] Liwiński, J. (2012). Rejestr cywilnych statków powietrznych. Ośrodek Informacji Naukowej, Technicznej i Ekonomicznej, ULC 2006-2012.

[29] Mansouri, S. H. & Bakhshan, Y. (2001). Studies of NO_x, CO, soot formation and oxidation from direct injection stratified-charge engine. Proc. Inst. Mech. Eng., Part. D, *Journal of Automobile Engineering*, *215*.

[30] Merkisz, J., Markowski, J., Pielecha, J. & Babiak M. (2010). Emission measurements of the AI-14RA aviation engine in stationary test and under real operating conditions of PZL-104 'Wilga' plane. *SAE Technical Paper Series*, 2010-01-1563.

[31] Merkisz, J., Markowski, J. & Pielecha J. (2009). Emission tests of small aircraft driven by piston engine in static conditions. Zakopane: International Conference „*Computer Systems Aided Science*, Industry and Transport" TRANSCOMP 2009.

[32] Merkisz, J., Markowski, J. & Pielecha J. (2009). Emission tests of the AI-14RA aircraft under real operating conditions of PZL-104 'Wilga' plane. *Combustion Engines*, *3*.

[33] Merkisz, J., Markowski, J. & Pielecha, J. (2010). Emission tests of the Cessna 152II aeroplane engine in stationary operating conditions. *Logistyka, 6.*

[34] Merkisz, J., Markowski, J. & Pielecha J. (2012). Evaluating piston engine aircraft emissions – structure of the stationary test. *Logistyka, 3.*

[35] Merkisz, J., Markowski, J. & Pielecha, J. (2011). The exhaust emission tests of a Zlin-142 M aircraft. *Combustion Engines, 3.*

[36] Merkisz, J., Markowski, J., Pielecha, J., Galant, M. & Karpiński, D. The exhaust emission verification for Zlin-142 M aircraft in stationary tests research. Zilina, Slovakia: International Conference on Air Transport "INAR 2013", University of Zilina, 71-75.

[37] Merkisz, J., Markowski, J., Pielecha, J. & Nykaza, A. (2012). Ocena emisyjności samolotu Zlin-142 M w warunkach lotu po kręgu. In T. Łodygowski, M. Ciałkowski (Eds.), *Rozwój Techniki, Technologii i Transportu w Lotnictwie.* Poznań, Wydawnictwo Politechniki Poznańskiej.

[38] Merkisz, J., Pielecha, J. & Radzimirski, S. (2009). *Pragmatyczne podstawy ochrony powietrza atmosferycznego w transporcie drogowym.* Poznań: Wydawnictwo Politechniki Poznańskiej.

[39] Merkisz, J., Pielecha, J. & Radzimirski, S. (2014). *New Trends in Emission Control in the European Union.* Switzerland: *Springer Tracts on Transportation and Traffic*, Vol. *1.*

[40] Merkisz, J. & Pielecha, J. (2008). *Reliability of particle emissions measurement on engine test stand.* Leipzig: 10 World Filtration Congress, Discover the Future of Filtration and Separation.

[41] Mongia, H. C. (2007). GE aviation low emissions combustion technology evolution. *SAE Technical Paper Series*, 2007-01-3924

[42] Moser, F. (2005). *Three different industries sharing the same powertrain technology – an opportunity for synergies?* Graz: 3rd AVL International Commercial Powertrain Conference.

[43] Nuccio, P. & Marzano, M. R. (2003). Performance improvement of two-stroke SI engines for motor-gliders and ultra-light aircraft by means of a GDI system. *SAE Technical Paper Series*, 2003-32-0002.

[44] Papagiannakis, R. G., Kotsiopoulos, P. N., Hountalas, D. T. & Yfantis, E. (2006). Single fuel research program comparative results of the use of JP-8 aviation fuel versus diesel fuel on a direct injection and indirect injection diesel engine. *SAE Technical Paper Series*, 2006-01-1673.

[45] Petersen, M. (2008). The legality of the EU's stand-alone approach to the climate impact of aviation: The express role given to the ICAO by the Kyoto protocol. *Blackwell Publishing Reciel*, *17* (2).

[46] Phuong, P. X. (2010). An experimental strategy for the manufacture of aviation fuel. *SAE Technical Paper Series*, 2010-01-1878.

[47] Regner, G., Johnson, D., Koszewnik, J., Dion, E., Redon, F. & Fromm, L. (2013). Modernizing the opposed piston, two stroke engine for clean, efficient transportation. *SAE Technical Paper Series*, 2013-26-0114.

[48] Rezy, B., Stuckas, K., Tucker, J. & Meyers, J. (1979). Concepts for reducing exhaust emissions and fuel consumption of the aircraft piston engine. *SAE Technical Paper Series*, 790605.

[49] Rindlisbacher, T. (2007). Aircraft piston engine emissions – summary report. Bern: Federal Office of Civil Aviation (FOCA), *Environmental Affairs*.

[50] Rodrigues Nerosky, L. A., Menezes, M. & Menezes Senna, J. C. S. (2007). Aviation and environment: certification and tendencies. *SAE Technical Paper Series*, 2007-01-2592.

[51] Schaefer, M. (2012). *Development of a forecast model for global air traffic emissions*. Dissertation zur Erlangung des Grades Doktor-Ingenieur der Fakultät für Maschinenbau der Ruhr-Universität, Bochum.

[52] Schumann, U. (2005). *Formation, properties and climatic effects of contrails*. Deutsches Zentrum fur Luft- und Raumfahrt, 2005.

[53] *Sensors SEMTECH-DS User Manual*, Revision 1.15, 2008.

[54] Shahinian, V. D. (2007). SENSORS tech-ct update application software for SEMTECH mobile emission analyzers. Ann Arbor: Sensors 4th Annual SUN (SEMTECH User Network) Conference.

[55] Slemr, F., Giehl, H. & Slemr, J. (1988). In-flight measurement of aircraft non-methane hydrocarbon emission indices. *Geophysical Research Letters*, *25*.

[56] Soban, D. S. & Upton, E. (2006). Towards electric aircraft: progress under the NASA URETI for aeropropulsion and power technology. *SAE Technical Paper Series*, 2006-01-3097.

[57] Sorokin, A., Vancassel, X. & Mirabel, P. (2003). Emission of ions and charged soot particles by aircraft engines. *Atmos. Chem. Phys.*, *3*.

[58] Taghavi, R. (2002). Propulsion education for general aviation aircraft. *SAE Technical Paper Series*, 2002-01-2926.

[59] Tilston, J., Larkman, J., Plohr, M., Döpelheuer, A., Lischer, T. & Zarzalis, N. (2003). Future engine cycle prediction and emission study. QinetiQ: Final Publishable Report.

[60] Turns, S. R. (1996). *An introduction to combustion concepts and applications*. New York: McGraw-Hill.
[61] Unger, N. (2011). Global climate impact of civil aviation for standard and desulfurized jet fuel. *Geophysical Research Letters*, *38*, L20803.
[62] Vespermann, J. & Wittmer, A. (2011). Financial, ecological and managerial impacts of emission trading schemes: the Case of Lufthansa. Business Strategy and the Environment. *Bus. Strat. Env.*, *20*.
[63] Wentz Jr., W. H. & Mohamed, A. S. (2008). Going green: what's in it for general aviation? *SAE Technical Paper Series*, 2008-01-2225.
[64] Wentz, Jr. W. H. & Mohamed, A. S. (2004). Preliminary design considerations for zero greenhouse gas emission airplanes. *SAE Technical Paper Series*, 2004-01-1803.

Chapter 3

INVESTIGATIONS INTO THE EXHAUST EMISSIONS FROM TURBOSHAFT ENGINES

Helicopters are used for naval, land and mountain rescue operations as well as for business and military purposes. The emission of pollutants from helicopters is not covered by exhaust gas emission standards, which inspired the Authors to carry out an assessment of emissions in exhaust gas from helicopter engines. On the basis of preliminary measurements an emission test was developed. It was designed in a manner reflecting operation conditions of such engines. Such a test can be performed during a pre-flight test, being a standard procedure used to assess helicopters' airworthiness. Emission measurements carried out during pre-flight tests coupled with emission test results made it possible to determine specific emissions of toxic compounds, based on which environmental performance of helicopter propulsion systems can be assessed.

3.1. INTRODUCTION

The need to reduce negative environmental impacts has always been one of the key determinants of technical development in all areas of transportation. The use and development of advanced technologies requires designers to constantly verify engines' operating conditions and the consequences of their impact on the natural environment. Transportation is considered a dynamically developing discipline mostly because of the reduced emissions of toxic compounds in exhaust gas. Emissions of toxic compounds in exhaust gas generated by aircraft propulsion systems continue to be a major threat. Particular attention is paid to carbon dioxide (as a measure of fuel consumption), and nitrogen oxides and particulate matter. Relative differences in nitrogen oxides and particulate matter formation mechanisms constitute a barrier in the development of contemporary internal combustion engines [1, 3, 4].

It is anticipated that in the forthcoming years the internal combustion engine will retain its dominating position in the aircraft industry. In the 21st century it seems that energy and environmental issues will be the key problems faced by mankind. The intensity of transportation-related negative impacts depends first of all on the number of vehicles and their efficient use. Therefore, thorough efforts have been taken for many years in order to minimize the said impacts. Growing concerns about the condition of the natural environment result in introducing increasingly strict standards on toxic emissions from internal combustion engines. As a result, legal provisions on the emission of pollutants from internal combustion engines used in various types of aircraft are now in place. Consequently, contemporary aviation engines are significantly different from those used less than twenty years ago, in terms of both design and technology.

Currently applicable limits require joint action of aviation engine manufacturers aimed at identifying new ways to cut emissions. Such actions take the form of joint programs, such as the EU's NEWAC, aiming at reducing carbon dioxide emissions by 6% and nitrogen oxides emission by 16%. The program is in keeping with the objectives of the Advisory Council for Aeronautics Research and Innovation in Europe (ACARE) that aim at a 50% reduction in carbon dioxide and 80% reduction in nitrogen oxides, as well as 50% reduction in noise by 2020, as compared to prescribed emission limits in effect in 2001 [2, 3, 4]. CLEAN SKY, another program implemented in EU member states, is a joint technology initiative aimed at supporting the development and deployment of new aircraft propulsion technologies that contribute to significant reduction in negative environmental impacts caused by air transportation. Also the International Air Transport Association (IATA) carries out a program dedicated to reducing aviation's negative environmental impacts by encouraging aircraft manufacturers to invest in new technologies, encouraging carriers to optimize flight schedules, and finally encouraging airports to invest in infrastructure ensuring maximized environmental care. As a result of these efforts, carbon dioxide emission in 2004 to 2007 was reduced by 44.5 million tons, equivalent to savings of 7.7 billion USD.

In EU countries it is estimated that aviation generates 3% of greenhouse gases. According to the Intergovernmental Panel on Climate Change, other means of transportation generate another 2% of emissions, as compared to 13% generated by road transportation. The level of greenhouse gas emissions is growing, a tendency that will surely be continued. From 1990 to 2009 the emission of aviation-originated greenhouse gases in EU countries grew at a constant rate of 4.3% p.a.

Statistical analyses investigating aircraft's effect on the environment as well as development works carried out as part of various projects largely concentrate on commercial and cargo airplanes and on turbine and turbofan engines used in their propulsion systems. However, no efforts have been made to introduce standards for turbine engines dedicated to turboprop and helicopter applications.

Therefore, this publication sets out to assess the impact of selected types of helicopters on the environment and to test this impact in a purpose-designed test procedure.

3.2. Share of Helicopters in Air Transportation Worldwide

Contemporary air transportation uses two types of aircraft: fixed-wing aircraft (i.e., airplanes) and rotary-wing aircraft (helicopters). The latter were first introduced during the 2nd World War. Their key advantage is that they are capable of performing aviation operations and tasks that are beyond the technical possibilities of airplanes. Helicopters are particularly useful in point-to-point transportation of people or goods, without the need to use costly airport infrastructure. As a result, helicopters are used across all spheres of human life and economy and are especially appreciated when other means of transportation prove ineffective, either for technical reasons or for time constraints. Helicopters are most commonly used in rescue or patrol operations, monitoring power infrastructure, communication with isolated places such as oil rigs, as well as in archeology, geology, agriculture, forestry and fishery. The above examples do not constitute an exhaustive list of all possible civil aviation applications. Furthermore, helicopters have a variety of military uses as well. Apart from tactical transportation of troops, weapons and provisions and from medical transportation, helicopters are also used in search and rescue operations, and in fighting ground targets and submarines. They can also be used for emitting electronic interferences (jamming) or for sweeping mines. Finally, they can serve as flying command posts. It is estimated that there are currently at least 58 000 helicopters in use globally, 41.5% of which are used by civil aviation and 58.5% by military aviation. In terms of absolute numbers, helicopters represent approximately 4% of civil aircraft and over 50% of military aircraft.

Registers of civil aircraft in the member states of the International Civil Aviation Organisation (ICAO) show that in 1990 the global fleet of civil helicopters amounted to 19 166 units. The said figure does not include former Soviet Union countries, where the number of registered aircraft was classified confidential. It is estimated that all USSR republics combined had approximately 5000 civil helicopters. If this figure is to be trusted, the total number of civil helicopters worldwide in 1990 grows to over 24 000 units. In relative terms however this figure constitutes a mere 4% of all aircraft in the world, most of which are airplanes.

The largest number of civil helicopters were registered in the USA (9454). Apart from the USSR, the other leading countries were Canada (1390) and Japan (1069). In Europe, the biggest helicopter fleets were owned by Great Britain (877), France (676) and Germany (457). With a fleet of 244 registered helicopters, Poland also held a significant place in the global ranking, but by the beginning of the 21st century the country's number shrank by a half (122). A geographic analysis of the said registers reveals that 45.8% of helicopters were used in North America, as compared to 30.3% in Europe (Tab. 3.1) [15].

Table 3.1. Civil helicopters in various parts of the world in 1990 [15]

Region	Helicopter type			Total
	piston	turbine, single engine	piston twin engine	
Europe	1819	1436	4056	7311
North America	4768	4967	1322	11 057
Central America and the Caribbean	24	45	8	77
South America	183	470	114	767
Australia and Oceania	543	413	80	1036
Africa	165	225	120	510
Asia	840	770	1603	3213
Middle East	23	81	84	188

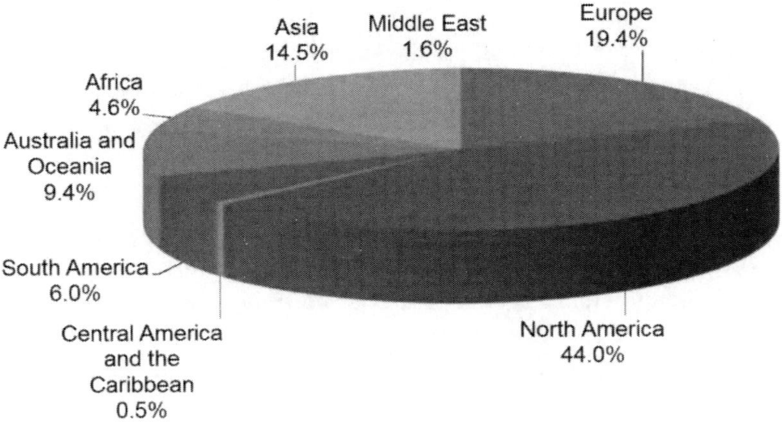

Figure 3.1. Relative share of world regions in the global number of civil helicopters in 2012 [15].

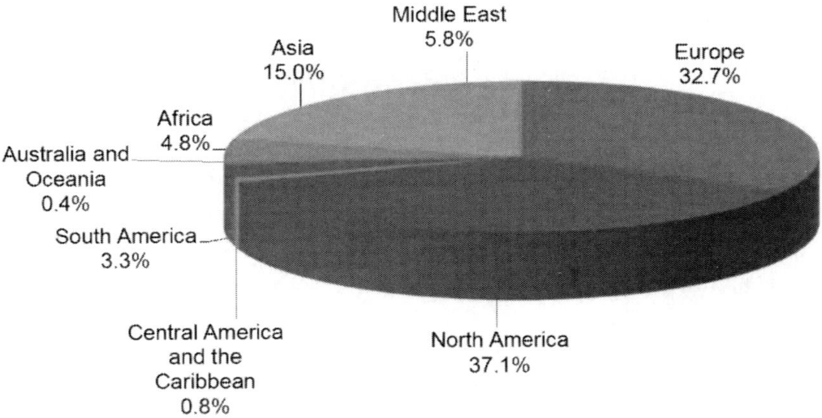

Figure 3.2. Relative share of world regions in the global number of military helicopters in 1991 [15].

An analysis of available databases [15] shows that the number of civil helicopters operated worldwide exceeds 42 000. The numbers for individual regions of the world are as follows: Europe – over 8000, North America – approx. 19 000, Central America and the Caribbean – approx. 200, South America – over 2500, Australia and Oceania – approx. 4000, Africa – approx. 2000, Asia – over 6000, Middle East – approx. 700. Each region's relative share of civil aircraft operated worldwide is presented in Figure 3.1. In Europe the biggest fleet is currently owned by France (1439), Great Britain (1292) and Germany (896). Given that in 1990 the number of civil helicopters in those countries amounted to 877 (Great Britain), 676 (France) and 457 (Germany) one can conclude that in 21 years the said numbers doubled in France and Germany and grew by approx. 50% in Great Britain.

The analysis of helicopters operated in Europe does not include Russia, where the number of currently registered civil helicopters amounts to 3017. The country was excluded from the analysis because of its peculiar geographical situation (split between Europe and Asia).

Attempting to assess the quantity of military helicopters one needs to realize that their number is constantly changing for natural reasons, and no army in the world has a habit of reporting its losses. Analyses presented in available literature [15] show that in 1991 the total number of helicopters operated by all armies of the world amounted to 34 125. Unlike in civil aviation, military helicopters represented a majority of all military aircraft (51%). The US army had the largest helicopter fleet (over 12 000 units) followed by former Soviet Union countries (6000 units or more). Most military helicopters were operated in North America (37.1%) and in Europe (32.7%) (Figure 3.2).

A characteristic feature of helicopter fleets is that – with the exception of the poorest countries – there are virtually no piston engine helicopters; while in civil aviation such helicopters represent as much as 34.6% of the entire fleet, they constitute a mere 3.8% of military fleets. The difference results from the nature of demands that must be met by contemporary army helicopters, the most important of which include technical capabilities that enable operation in all weather conditions on more than 95% of the world's surface, resistance to mechanical and ballistic damage, resistance to and crew protection against the consequences of hard landing, simple operation, compatibility with various types of fuel, etc.

Not unlike for civil helicopters, the figures on the number of military helicopters currently in use are not up to date, but the relative shares of various parts of the world calculated on that basis remain similar.

The estimated share of helicopters in the global number of aircraft amounts to approximately 4%. The corresponding value for military aircraft is close to 50%. 95% of military helicopters are propelled with turbine engines, appreciated for concentration of power as well as for their reliability across all climatic conditions. However, engines of this kind are characterized by high instantaneous fuel consumption, twice greater than for piston engines. Therefore, their energy efficiency is lower which translates into high fuel consumption per hour. In its turn, fuel consumption translates into emission of toxic compounds in exhaust gas. Global economic development creates favorable conditions for a growth in the number of helicopters, which is why it seems necessary to evaluate emissions from helicopter turbine engines and their effect on the environment.

Such an evaluation may contribute to an introduction of limits of toxic compounds in exhaust gas, constituting a powerful tool stimulating the development of this type of engines. Current analyses and future studies into emissions of toxic compounds from helicopter propulsion systems may provide a basis for assessing emissions of helicopters, both in pre-trial tests and in actual operation conditions. An assessment of such actual operation conditions will help develop a test whose static phases and their relative coefficients will make it possible to represent the conditions affecting engine loads in in-flight conditions. Such tests could be used for assessing the gravity of environmental impacts caused by helicopter propulsion systems. In the case of helicopters operated in a battlefield situation, environmental impacts in the form of toxic compounds in exhaust gas are of marginal importance.

3.3. Testing Methodology

Determining the values of emissions from helicopter propulsion systems and thus identifying helicopters' impact on the natural environment requires a number of tasks to be completed:

- determination of gaseous compounds in exhaust gas emitted from helicopter engines during pre-flight tests,
- analysis of engine load histograms in in-flight conditions for two helicopter types,
- analysis of exhaust gas emission in in-flight conditions,
- development of a test determining the emission of pollutants in exhaust gas generated by propulsion systems of transport helicopters.

The issue tackled by the Authors requires development of a testing methodology dedicated to emissions from helicopter propulsion systems in in-flight conditions. Transport helicopters undergo pre-flight tests in order to verify the performance of all their instruments and systems. An analysis of helicopters' designs and their operational capabilities was carried out in order to develop a method of measuring the concentrations of toxic compounds in exhaust gas. Concentrations of carbon dioxide, carbon monoxide, hydrocarbons and nitrogen oxides in exhaust gas were measured during pre-flight tests of available helicopters. For helicopters fitted with flight recorders, concentrations of toxic emissions were recorded simultaneously with engine operation parameters during pre-flight tests. Subsequently, all recorded data were analyzed comparatively, based on which changes in toxic compound emissions were correlated with changes in engine operation conditions. Once the relationships between toxic compounds' concentrations on the one hand and helicopter and engine operation parameters on the other became known, an analysis was carried out to calculate emission rates of toxic compounds at various engine load values [5, 14].

Information on toxic emission rates coupled with information on accompanying engine loads was used to assess toxic emissions in exhaust gas in actual operation conditions. The assessment was carried out on the basis of data from in-flight recorders, subjected to an analysis of engine load statistic frequency during the flight. A comparison of load histograms with corresponding toxic emission rates made it possible to assess in-flight emissions from helicopters. It needs to be remarked that the analysis is burdened with a certain degree of inaccuracy. Its verification requires a series of in-flight measurements of toxic emissions, which are currently unfeasible for a number of reasons beyond the authors' control.

Table 3.2. Tested helicopters

Helicopter type		**No. of engines**	**Engine**	**Power**
Mi-2		2	GTD-350	298 kW
Mi-17		2	TW3-117MT	1546 kW
PZL-W-3 Sokół		2	PZL-10W	662 kW

Helicopter type		No. of engines	Engine	Power
PZL- -SW-4 Puszczyk		1	Rolls Royce (Allison) 250-C20R/2	336 kW

http://pl.wikipedia.org/wiki/Plik:Mi-2_Plus_Lotnicze_Pogotowie_Ratunkowe.jpg.
http://en.wikipedia.org/wiki/File:KazakhstanMi-8MT2000(DF-SD-01-06442).jpg.
http://pl.wikipedia.org/wiki/Plik:Topr.jpg.
http://pl.wikipedia.org/wiki/Plik:PZL_SW-4_Solo_RAUS_OPH.jpg.

The tests involved five helicopters, all of which are classified as transportation aircraft and represent their respective helicopter categories. These were: Mi-2 (2 units), Mi-17, W-3 Sokół and SW-4 Puszczyk (Tab. 3.2).

Engines Used in Propulsion Systems of the Tested Helicopters

GTD-350 turbine engine

The GTD-350 engine was designed as an aviation engine dedicated to the Mi-2. It is fitted with two impellers and a separate power turbine. An overview of its key components and its working principle are presented in Figure 3.3. Air is supplied to GTD-350 via a compressor with seven axial stages and one centrifugal stage, and is subsequently fed into a can combustor. Fuel is delivered to the chamber by an injector valve. Fuel combustion generates a growth in the thermodynamic parameters of the working medium, which – once directed to the high stage turbine – drives the compressor. Subsequently, the working medium is directed to the power turbine that drives the propeller shaft via transmission gear. Such a design is characteristic for the placement of the combustion chamber downstream of the turbines. As a result, it is possible to use shorter impeller shafts and thus high engine speeds. Such a solution had one more significant advantage: reduced engine mass and improved economical performance. Mi-2's propulsion system consists of two GTD-350s that propel the final drive, and – indirectly – the main and auxiliary rotor and other units coupled with the transmission [3].

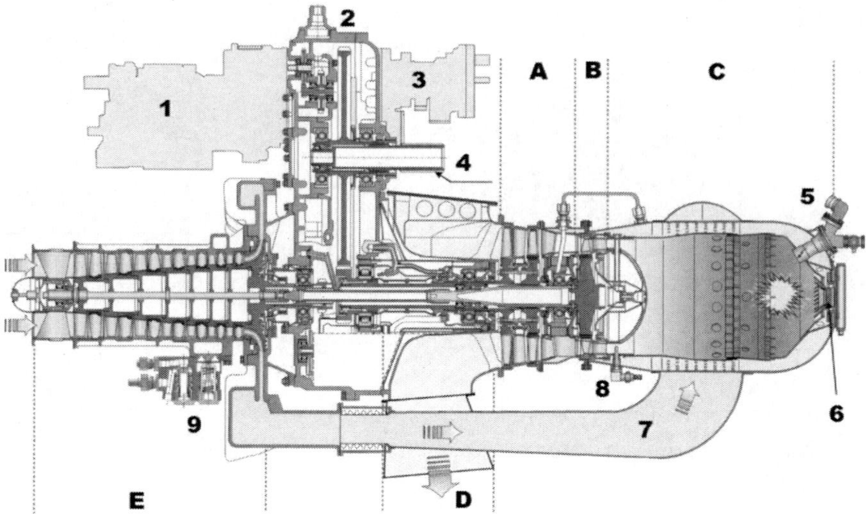

Figure 3.3. GTD-350 engine overview [3]: A – power turbine, B – compressor turbine, C – combustion chamber, D – exhaust gas outlet, E – compressor, 1 – regulator pump, 2 – gearbox, 3 – power turbine speed governor, 4 – power take-off, 5 – igniter plug, 6 – injector valve, 7 – air tube, 8 – thermocouple, 9 – wastegate.

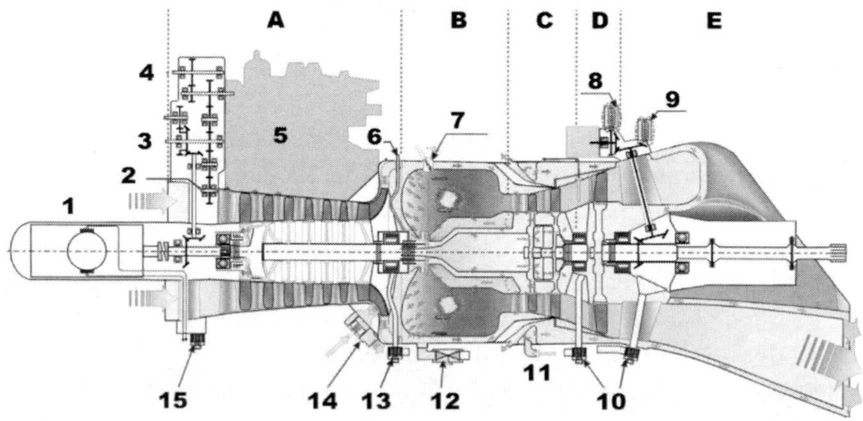

Figure 3.4. Schematics of silnika PZL-10W [3]: A – compressor, B – combustion chamber, C – compressor turbine, D – power turbine, E – power take-off and exhaust gas outlet, 1 – starter, 2 – oil pumps drive, 3 – manual turbocompressor impeller turning socket, 4 – generator drive, 5 – regulator pump, 6 – fuel supply, 7 – igniter plug, 8, 9 – speed sensors, 10 – magnetic plugs, 11 – **from helicopter's fan,** 12 – drip valve, 13 – magnetic plug, 14 – wastegate, 15 – magnetic plug.

PZL-10W turbine engine

The PZL-10W engine fitted with a free-wheeling turbine is used in the W-3 Sokół. The power unit consists of two PZL-10W engines and one WR-3 final drive, complete with a number of units and devices ensuring failure-free operation across various climatic conditions. Power generated by the engines is transmitted to the final drive which directly drives the main rotor and indirectly – via intermediate and tail gearboxes – the tail rotor. In an emergency, the W-3 can continue flying on one engine, although some operation ranges will be unavailable [3]. An overview of the engine's key components and its working principle are presented in Figure 3.4.

The PZL-10W is a classical turbine engine with a free-wheeling turbine fitted with an inlet assembly with an embedded ancillary units gearbox. An electric starter with a one-way clutch is installed upstream of the inlet, in the compressor's central axis. The engine's compressor features six axial stages and one (last) centrifugal stage. The other functional units of the engine are: combustion chamber unit, compressor-driving turbines unit, propulsion unit (turbine & propeller shaft), and finally exhaust gas outlet. Air is sucked in by the compressor via a heated inlet duct. Adequately compressed air is directed to the ring combustion chamber, where it is mixed with fuel sprayed by the rotating injector valve. The resulting fuel and air mixture is ignited by an igniting device. The resulting hot gases from the combustion chamber pass through the turbine guide and reach 1st and 2nd stage blades of the compressor-driving turbine, and then the power turbine. After the energy is released, exhaust gas is directed outside the helicopter via an outlet nozzle. The power required to drive the regulator pump and oil pump units is taken off the compressor's shaft via a gearbox. The helicopter's final drive used to power the main rotor and the tail rotor is propelled by means of the propeller shaft, coupled directly with the power turbine. The power turbine speed governor is fitted on the power turbine cylinder.

The engine comes with an anti-ice system protecting the compressor's inlet. In terms of the heating medium, the system is a mixed one. In one mode, the heating medium is hot air taken downstream of the compressor (open type) and in the other mode the heating medium is hot oil from the engine lubrication system (closed type) [3].

Rolls-Royce 250-C20 R/2 turbine engine

The SW-4 Puszczyk is powered by one Rolls-Royce 250-C20R/2 turbine engine with 450 kW of take-off power and a pneumatic-hydromechanical fuel control system (Figure 3.5). This engine's design and working principle are similar to those of the GTD 350, with a number of minor differences in the designs of individual units (Figure 3.6).

Figure 3.5. Rolls-Royce 250-C20R/2 engine fitted on the SW-4 Puszczyk.

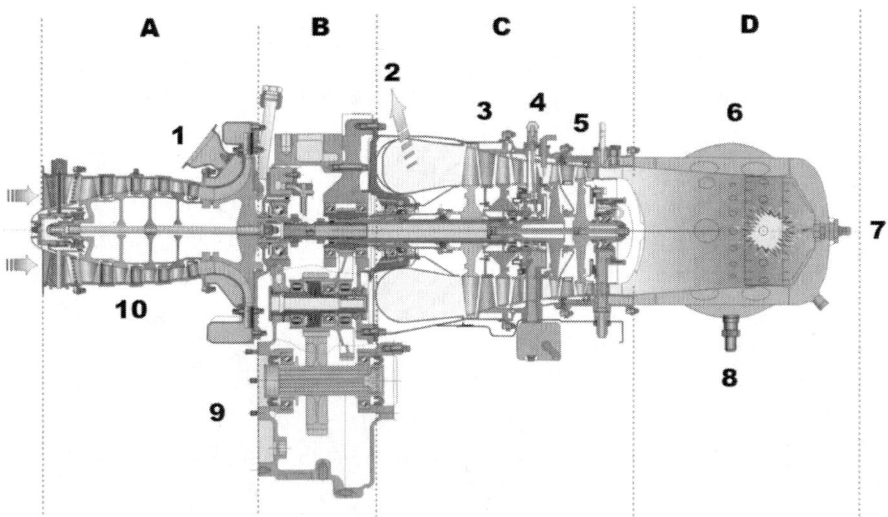

Figure 3.6. Rolls-Royce 250-C20R/2 engine overview [3]: A – compressor assembly, B – gearbox assembly, C – turbine assembly, D – combustion assembly, 1 – wastegate base, 2 – exhaust gas outlet, 3– power turbine, 4 – thermocouple, 5 – compressor turbine, 6 – combustion chamber, 7 – fuel injector, 8 – drip valve, 9 – power take-off, 10 – compressor.

The engine has a can combustor; its compressor has four axial stages and one centrifugal stage; the turbines unit features two stages in the compressor-driving turbine and two stages in the power turbine. The engine power is transmitted to the two-stage final drive and the single-stage rear drive via the one-way clutch shaft. The engine has a pneumatic-hydromechanical fuel control system. As a result, the engine can continue to operate despite an electric power supply failure. However, in such a situation the controls of the propulsion system, the electromechanical engine speed control system and the ignition system will not work (no engine start possible). The engine is fitted with an anti-ice system relying on hot air collected from the compressor's wastegate [3].

TW3-117 turbine engine

The propulsion system of the Mi-17 features two turbine engines TW3-117 dedicated to helicopters (Figure 3.7). The TW3-117's design and working principle are similar to those of the PZL-10W, with a number of minor differences. The TW3-117 features several main functional units, such as: air intake, two-stage axial compressor, ring combustion chamber, two-stage compressor turbine, two-stage power turbine and exhaust outlet. The main propeller shaft passes through that outlet. The engine has a gearbox for ancillary units driven by the first stage compressor shaft in the engine's inlet section.

Figure 3.7. TW3-117 engine with cross-sections showing its internal components [16].

Table 3.3. Characteristics of the analyzed helicopters' engines

Engine	Power [kW]	Instantaneous fuel consumption [kg/kWh]	Air flow rate [kg/s]	Compression ratio [–]	Engine weight [kg]
GTD-350	294	0.475	2.2	6.05	135
PZL-10W	660	0.408	4.5	7.0	150
RR 250-C20R	380	0.465	1.7	7.9	78
TW3-117	1640	0.410	8.1	6.6	295

Apart from the two TW3-117s, the propulsion system of the Mi-17 is fitted with AI-9W, an auxiliary system for starting up the main engines. The auxiliary AI-9W engine is fitted in the last compartment of the engine assembly shield. It is positioned crosswise to the helicopter's fuselage, at an inclination of 3°30'. The weight of this low-power turbine engine is 70 kg. The engine is started by means of on-board batteries or from an external power source.

Key specifications of the engines related to a number of parameters required to analyze toxic emissions are presented in Table 3.3.

3.4. Study of Exhaust Emissions from Helicopters in Pre-Flight Tests

The first stage of the experiment aimed at accomplishing the objective of the project was to measure the concentrations of toxic compounds in exhaust gas from the engines used in the propulsion systems of the selected helicopters. The measurements were carried out during pre-flight tests constituting part of the helicopters' standard operating procedure. The purpose of such a test is to verify the helicopter's condition and to assess its airworthiness. The test includes checks of all on-board devices, propulsion system and control systems. Sample standard pre-flight test procedures for the selected helicopters are presented in Tables 3.4–3.7. Pre-flight tests make it possible to determine engine loads generated during the test and their respective durations, thus making it possible to assess the time density of loads. Engine loads can be assessed to a certain degree of accuracy based on characteristics representing interrelations between main rotor speed and engine loads at given values of atmospheric pressure and air temperature. An analysis of pre-flight test procedures confirmed the existence of steady states in helicopter engine loads whose values are close to the maximum load. Therefore, operation conditions during pre-flight tests were considered adequate for measuring the concentration of toxic compounds in exhaust gas without additionally increasing the aircraft's operation time.

Emissions of toxic compounds in exhaust gas were analyzed in actual operation conditions, during pre-flight tests. To do so, it was necessary to install an exhaust gas sampling device on the helicopter, close to the engine's exhaust gas outlet but without compromising the engine's operation. A number of simple structural elements were used for that purpose (Figure 3.8). The exhaust gas sampling system in the form of a duct was connected to the analyzer via an open window in the helicopter's cargo space.

Table 3.4. Pre-flight test of the Mi-2 [3]

Start of procedure	End of procedure	Performed procedure	Engine parameters
0'00"	0'40"	start-up of the first engine	transitional load ranges until the minimum speed is achieved
0'45"	1'25"	start-up of the first engine; in the meantime the first engine is warmed up; after the second engine has started, external power supply is disconnected	minimum engine speed, $n_{TS} = 57\%\ n_{max}$
1'25"	3'00"	warming up the engines, checking airframe devices, avionics, etc.	minimum engine speed, $n_{TS} = 57\%\ n_{max}$
3'00"	3'15"	engine speed increases up to the take-off power	engine speed, $n_{TS} = 96\%\ n_{max}$, power $N_e = 294.2$ kW
3'15"	8'00"	hovering, checking the helicopter's manoeuvrability, operation of avionics	engine speed, $n_{TS} = 96\%\ n_{max}$, power $N_e = 294.2$ kW
8'00"	10'00"	landing and engine cooling, preparation for helicopter shutdown	minimum engine speed, $n_{TS} = 57\%\ n_{max}$

Table 3.5. Pre-flight test of the Mi-17 [3]

Start of procedure	End of procedure	Performed procedure	Engine parameters
0’00”	2’00”	start-up of the start-up assembly (AI-9W engine), warming up, stabilization at idling speed	warming up until the target exhaust gas temperature is reached; warm-up time: 1 minute; temperature at idling speed: max 720°C; maximum operation time: 30 minutes
2’00”	10’00”	start-up of the TW3-117 engine, warming up of the first engine, start-up of the second engine, disconnection of external power supply	AI-9W exhaust gas generator speed approx. 73% n_{max}, exhaust gas temperature 780°C; target power turbine and 1st engine speed: 40–55% n_{max}; engine warm-up time: 1 minute at idling speed
10’00”	13’00”	warming up the engines, checking airframe devices, avionics	AI-9W current generator range toggle (on) until 30 minutes less start-up time; check voltage and current consumption in the current generator mode
13’00”	13’30”	engine speeds increases up to the take-off power	engine power turbine speed 91–97% n_{max}, power N_e = 1640 kW
13’30”	18’30”	hovering, checking the helicopter’s manoeuvrability, operation of avionics	procedures as per pre-flight manual

Start of procedure	End of procedure	Performed procedure	Engine parameters
18'30"	21'00"	landing and engine cooling, preparation for helicopter shutdown	switch off the engines when cooled down; measure the time from switch-off to stop

Table 3.6. Pre-flight test of the PZL W-3 Sokół

Start of procedure	End of procedure	Performed procedure	Engine parameters
0'00"	0'40"	start-up of the first engine	transitional load ranges until the minimum speed is achieved
0'45"	1'25"	start-up of the first engine; in the meantime the first engine is warmed up; after the second engine has started, external power supply is disconnected	minimum engine speed, $n_{TS} = 65–68\%\ n_{max}$
1'25"	3'00"	warming up the engines, checking airframe devices, avionics	minimum engine speed, $n_{TS} = 65–68\%\ n_{max}$
3'00"	10'00"	checking key systems, engine speeds increases up to the take-off power	the following conditions for increasing engine speed must be met prior to proceeding, engine speed $n_{TS} = 81.5–83.5\%\ n_{max}$, exhaust gas temperature $T_{exh} = 485–505°C$, shaft torque $M_o = 48.9–51.9\%\ M_{o\,max}$
10'00"	17'00"	hovering, checking the helicopter's manoeuvrability, operation of avionics	engine speed, $n_{TS} = 91\%\ n_{max}$

Table 3.6. (Continued)

Start of procedure	End of procedure	Performed procedure	Engine parameters
17'00"	20'00"	landing and engine cooling, preparation for helicopter shutdown	minimum engine speed, $n_{TS} = 65–68\%\ n_{max}$

Table 3.7. Pre-flight test of the PZL SW-4 Puszczyk

Start of procedure	End of procedure	Performed procedure	Engine parameters
0'00"	0'40"	start-up of the engine; after the engine has started, external power supply is disconnected	transitional load ranges until the minimum speed is achieved
0'45"	1'25"	warming up the engines, checking airframe devices, avionics	minimum engine speed, $n_{TS} = 57\%\ n_{max}$
1'25"	3'00"	engine speeds increases up to the take-off power	engine speed, $n_{TS} = 96\%\ n_{max}$, power $N_e = 380$ kW
3'00"	8'00"	available power test (helicopter stationary)	engine speed, $n_{TS} = 96\%\ n_{max}$, power $N_e = 380$ kW
8'00"	10'00"	landing and engine cooling, preparation for helicopter shutdown	minimum engine speed, $n_{TS} = 57\%\ n_{max}$

The emission of toxic compounds was determined on the basis of carbon monoxide, hydrocarbons, nitrogen oxides and carbon dioxide concentrations, measured with Semtech-DS, a portable exhaust gas analyzer by Sensors. Engines were started in the cold start mode, which means that they were kept at ambient temperature for approximately 24 hours before the start. The measurements' duration depended on the duration of the test itself and usually took 20 minutes.

During the helicopters' pre-flight test, operation parameters of their engines were recorded by means of on-board flight recorders. Changes in the engines' operation parameters recorded during the test were then compared against simultaneously recorded changes in toxic compound concentrations. In this way, respective concentration values were linked to engine loads during the test. Subsequently, an analysis of toxic emissions during the pre-flight test was carried out.

The results of toxic compounds concentrations in exhaust gas are presented as measured values of carbon monoxide, hydrocarbons, nitrogen oxides and carbon dioxide recorded over several minutes of measurement (Figs. 3.9–3.13). Measurements were carried out for all selected helicopters during their respective pre-flight tests. Individual tests may differ in terms of duration from the accepted procedure presented in the guidelines applicable to a given helicopter type. However, engine loads during the tests were strictly respected at all times.

Figure 3.8. Installation of the exhaust gas sampling system.

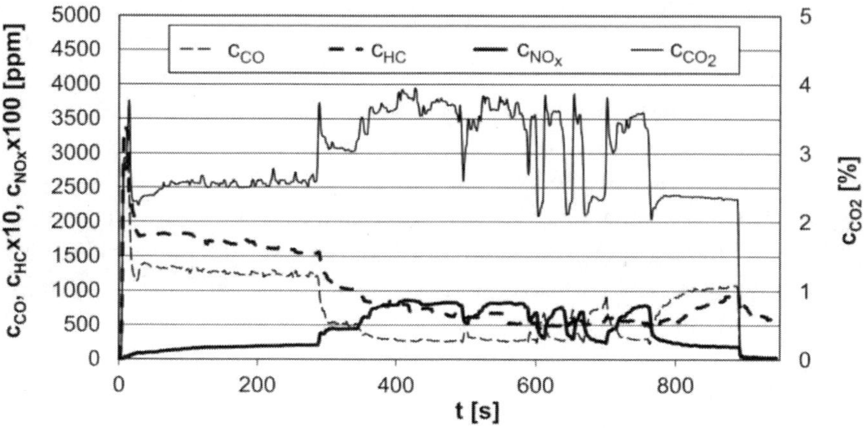

Figure 3.9. Concentrations of toxic compounds in exhaust gas from the engine of Mi-2 A.

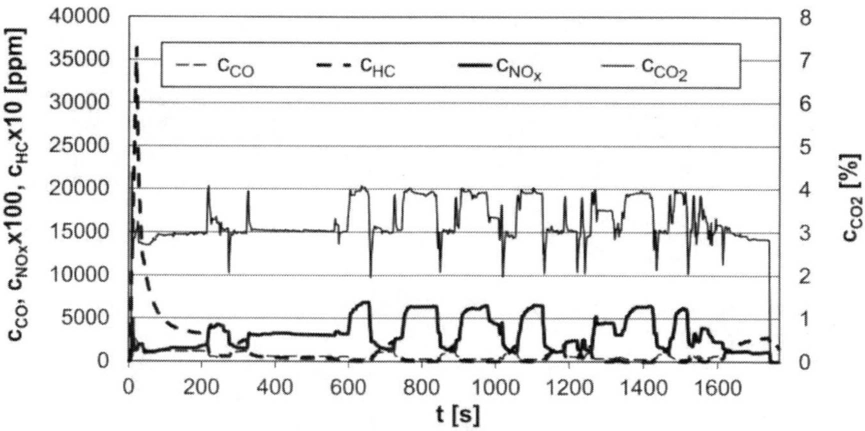

Figure 3.10. Concentrations of toxic compounds in exhaust gas from the engine of Mi-2 B.

The beginning of the engine start-up phase is a characteristic point in the charts presenting the concentration of toxic compounds in exhaust gas from turbine engines. This phase of the test shows only a high concentration of hydrocarbons, directly linked to fuel supply to the combustion chamber. Subsequently, upon ignition in the chamber, the concentrations of carbon monoxide, carbon dioxide and hydrocarbons grow immediately, as a consequence of the combustion process, the efficiency of which in the initial phase is low. The concentrations of carbon monoxide and hydrocarbons are quickly reduced as the engine warms up.

Higher engine temperature contributes to greater nitrogen oxides concentration in exhaust gas. During the test, nitrogen oxides and carbon dioxide concentrations at a warm engine are similar in nature, as both depend on engine loads. As the load grows, more fuel is supplied to the combustion chamber, which increases the temperature in the flame zone and inside the chamber, which in its turn is conducive to the formation of nitrogen oxides. Concentration values of toxic compounds in exhaust gas vary depending on the engine type used for propulsion. Engines used in the propulsion systems of the selected helicopters can be divided into two types of design. The first type are turbine helicopter engines with a can combustor. Such engines are used in Mi-2 and PZL SW-4 Puszczyk. The other design is a turbine engine with a ring combustion chamber, as used in Mi-17 and PZL W-3 Sokół. For can combustor engines, carbon dioxide concentrations stay in the range of 2–4% for the GTD 350 engines and in the range of 1.5–2.0% for the RR250-C20R/2 engine. Nitrogen oxides concentrations are up to 70 ppm. As regards carbon monoxide concentration in the case of the GTD 350 engines, it amounts to 1400 ppm at low loads and approximately 400 ppm at the maximum load. For the RR250-C20R/2 engine, the concentration of carbon monoxide amounts to approx. 900 ppm at low loads, and – just like for the GTD 350s – approximately 400 ppm at the maximum load. In the case of ring combustion chamber turbine engines, such as the PZL-10W and the TW3-117, carbon dioxide concentrations range from 1.5 to 3.5%. Nitrogen oxides concentration is approximately 25 ppm at partial loads and approx. 70 ppm at the maximum load. Nitrogen oxides concentration below 70 ppm at the maximum load is linked to the specific nature of the combustion process. In turbine engines, combustion is a constant process carried out at constant pressure and temperature values in the combustion chamber. Low changeability of the flame is conducive to low emission of nitrogen oxides. Once the engine has warmed up, carbon monoxide concentrations do not exceed 40 ppm. Hydrocarbon concentrations at warm engine were similar in all tests and amounted to approximately 50 ppm. The greatest amounts of hydrocarbons are generated during engine start-ups and switch-offs. The underlying reason are the initial difficulties with the air and fuel mixture preparation during the start-up and steaming of heavy hydrocarbons once fuel is no longer supplied and the flame is extinguished in the combustion chamber.

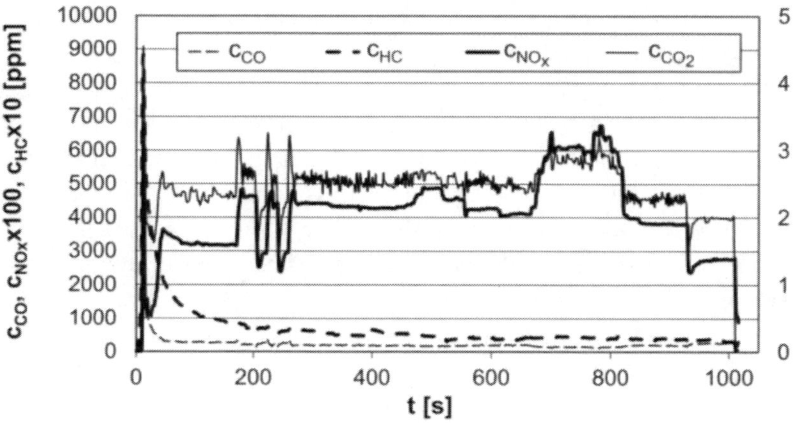

Figure 3.11. Concentrations of toxic compounds in exhaust gas from the engine of Mi-17.

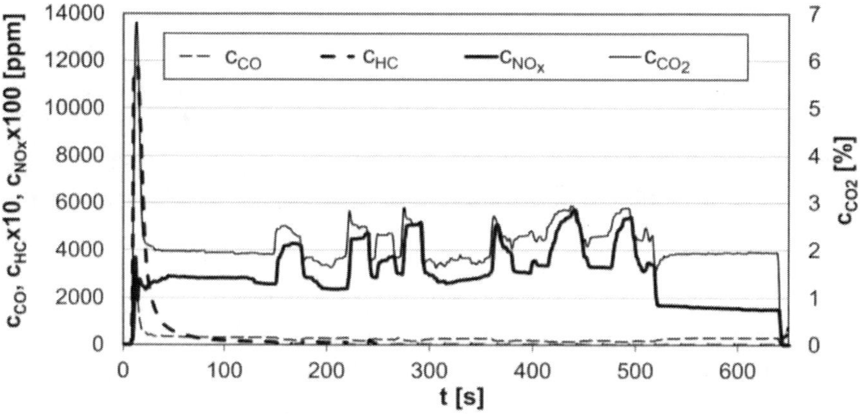

Figure 3.12. Concentrations of toxic compounds in exhaust gas from the engine of W-3 Sokół.

Out of all analyzed aircraft, two were fitted with on-board flight recorders: W-3 Sokół and SW-4 Puszczyk. Such recorders are used to register the helicopter's in-flight operation parameters. Some of the recorded values are related to engine operation parameters as well. Figure 3.14 shows a sample record of selected engine operation values registered by the on-board recorder of the W-3 Sokół during its pre-flight test. The engine load analysis took into account the exhaust gas generator turbine speed of both engines (N1_1 and N1_2), expressed as a percentage of the maximum speed, as well as the temperature downstream of the turbine (TOT_1 and TOT_2) shown in degrees Celsius. The sample chart below presents changes in instantaneous values of operation parameters of the PZL W-3 Sokół during its pre-flight test. The first part of the chart corresponds to the start-up of the helicopter's instrumentation and verification of on-board systems. Start-up of the engines is shown in the central part of the chart (engine 1 followed by engine 2). Based on the recorded changes engine load time shares were determined (Figure 3.15). The engines work at approximately 20% of the maximum load for 24% of the pre-flight test time, as compared to 43% time at loads ranging from 65 to 75% of the maximum load, and to 33% time at the maximum load.

A similar analysis of pre-flight data was carried out for PZL SW-4 Puszczyk. First, changes in recorded operation parameters of the engine recorded during the test were presented (Figure 3.16), based on which engine load time shares were determined (Figure 3.17). The time-based analysis of engine loads indicates that for 80% of the test time SW-4 Puszczyk operated at approximately 95% of the maximum engine load, as compared to 7% time at the minimum load. The remaining 13% time corresponds to operation at approximately 60% of the maximum load.

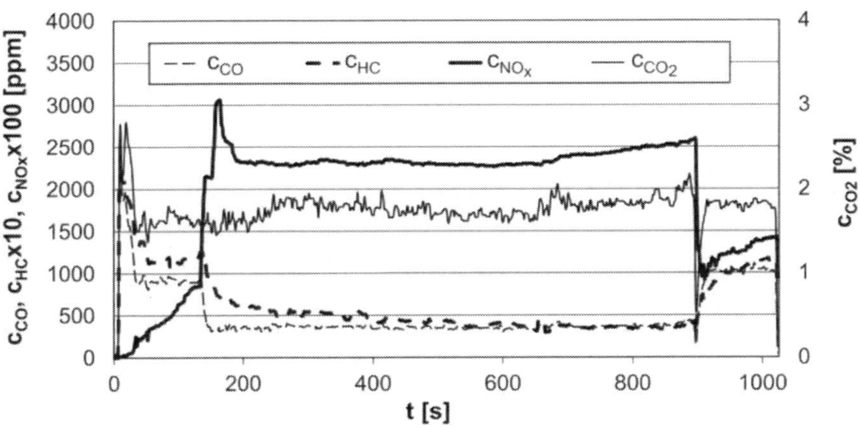

Figure 3.13. Concentrations of toxic compounds in exhaust gas from the engine of SW-4 Puszczyk.

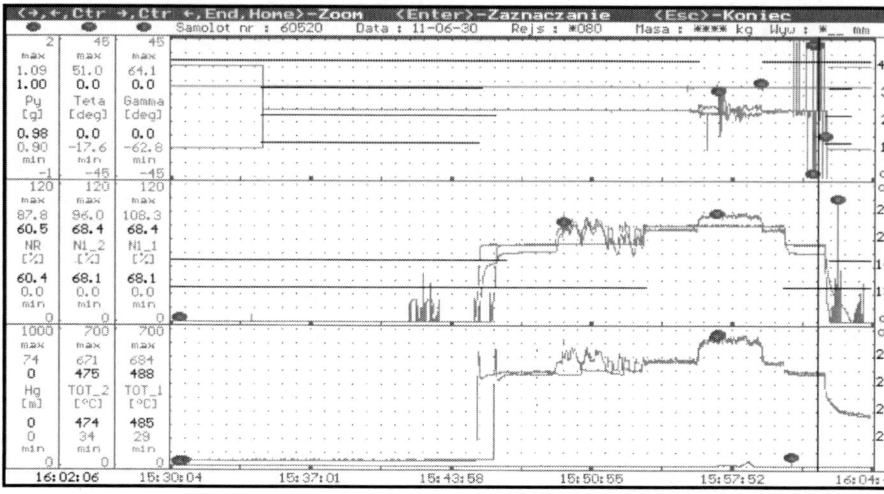

Figure 3.14. Example of engine operation parameters recorded during a pre-flight test of W-3 Sokół (source: on-board flight recorder).

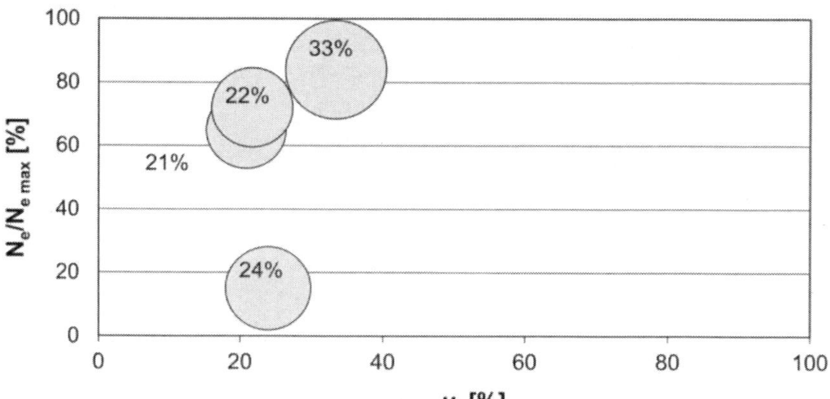

Figure 3.15. Propulsion system loads during a pre-flight test of W-3 Sokół.

By comparing the changes in the concentrations of toxic compounds in exhaust gas against changes in engine operation parameters recorded by the on-board flight recorder it is possible to determine concentration values typical of given engine loads. Such information can be derived only for the PZL-10W engine used in the propulsion system of the PZL W-3 Sokół (Figure 3.18) and for the RR 250-C20R/2 engine used in the propulsion system of PZL SW-4 Puszczyk (Figure 3.19).

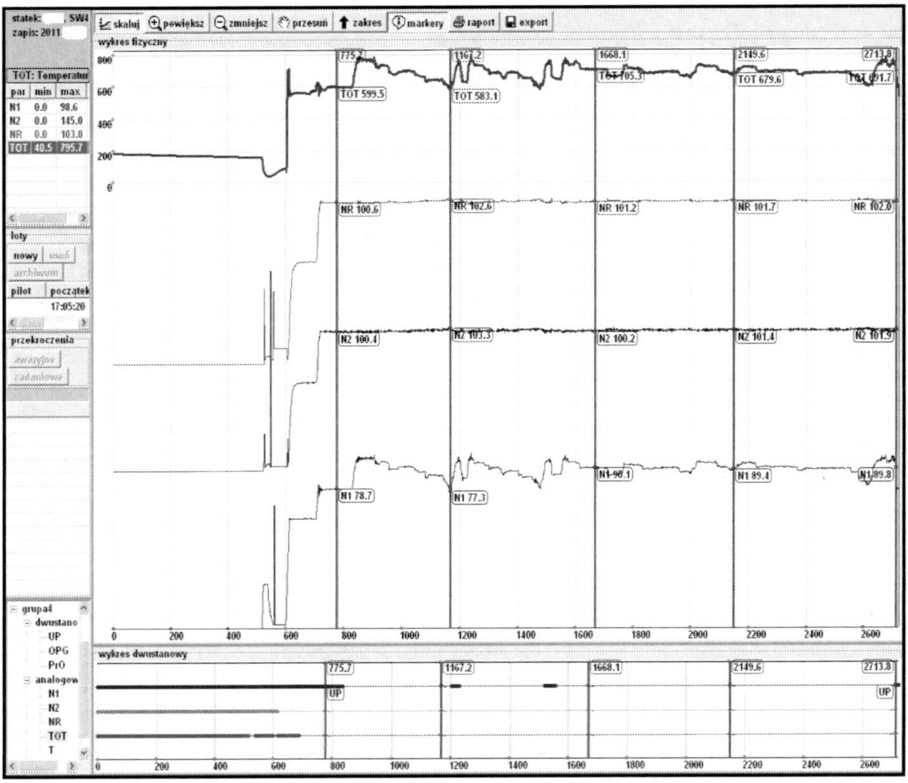

Figure 3.16. Example of engine operation parameters recorded during a pre-flight test of SW-4 Puszczyk (source: on-board flight recorder).

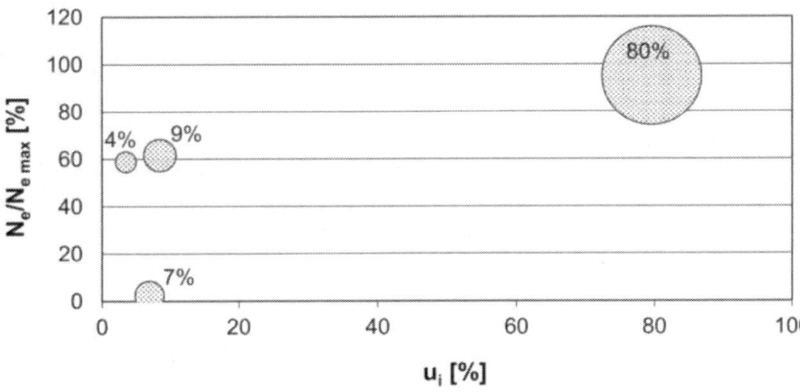

Figure 3.17. Propulsion system loads during a pre-flight test of W-3 Sokół.

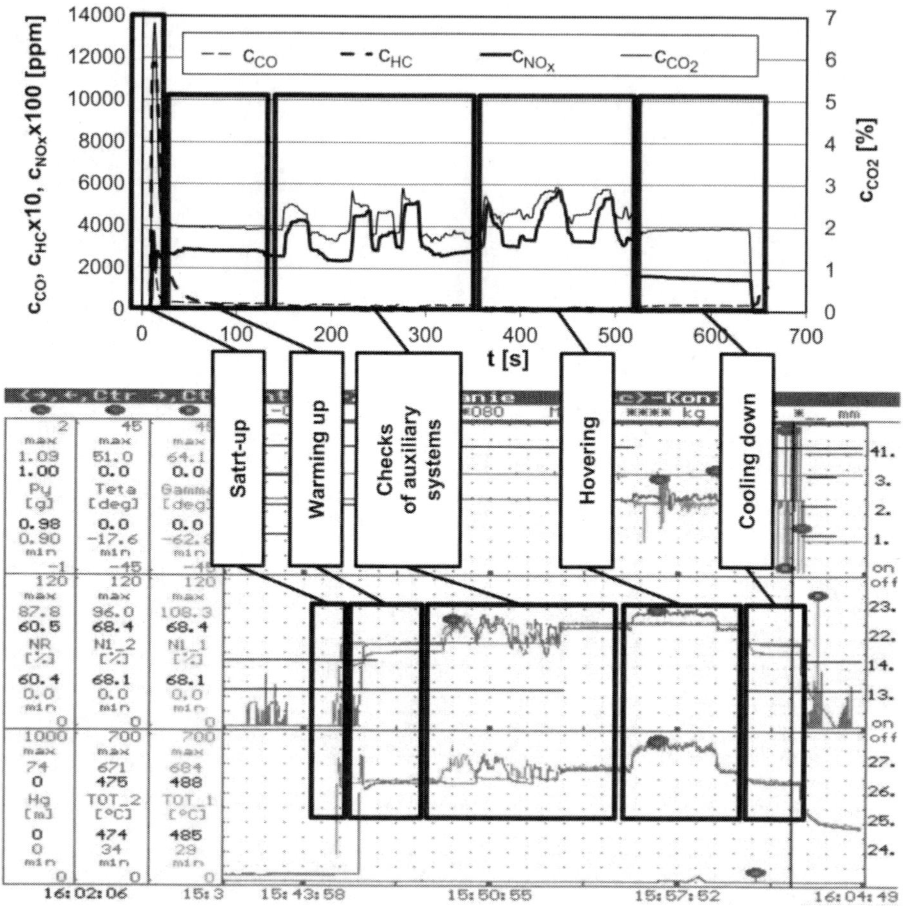

Figure 3.18. Comparison of changes in toxic compounds' concentrations against changes in engine parameters (PZL W-3 Sokół).

A comparison of toxic compounds' concentrations against engine loads made it possible to determine the concentrations of carbon dioxide, carbon monoxide, hydrocarbons and nitrogen oxides for different engine load points of both helicopters, i.e., PZL W-3 Sokół and PZL SW-4 Puszczyk (Table 3.8).

Neither Mi-2 nor Mi-17 were fitted with on-board flight recorders, which is why their respective engine loads were assessed on the basis of the pre-flight test procedure coupled with measured concentrations of toxic compounds in exhaust gas. Just like in the case of changes in toxic compounds' concentrations in exhaust gas from the engines of PZL W-3 Sokół and PZL SW-4 Puszczyk, different engine load states can be recognized. A comparison against the guidelines for engine operation parameters in the pre-flight test allows one to associate engine loads with corresponding concentrations of toxic compounds in exhaust gas (Table 3.9). In order to determine emission from each engine, concentrations of toxic compounds in exhaust gas were used as the starting point for the analysis. Determination of toxic emissions in exhaust gas requires one to calculate the mass of exhaust gas generated by the engine at each operation point. During the measurement of toxic compounds' concentrations in the pre-flight test it was not possible to measure air or fuel consumption. Likewise, it was impossible to directly measure exhaust gas output. Therefore, an attempt was taken at estimating the above values.

Table 3.8. Concentrations of CO_2, CO, HC, NO_x for selected engine loads of PZL W-3 Sokół and PZL SW-4 Puszczyk

Helicopter/engine	$N_e/N_{e\,max}$ [%]	c_{CO_2} [%]	c_{CO} [ppm]	c_{HC} [ppm]	c_{NOx} [ppm]
PZL W-3 Sokół/PZL-10W	15	2.0	500	40	30
	65	2.4	400	35	38
	72	2.5	380	34	50
	84	2.9	370	33	58
PZL SW-4 Puszczyk/RR250.C20R/2	15	1.4	900	115	5
	60	1.7	350	75	23
	95	2.0	345	35	25

Table 3.9. Concentrations of CO_2, CO, HC, NO_x for selected engine loads (Mi-2 and Mi-17)

Helicopter/engine	$N_e/N_{e\,max}$ [%]	c_{CO_2} [%]	c_{CO} [ppm]	c_{HC} [ppm]	c_{NOx} [ppm]
Mi-2 A/GTD 350	15	2.5	1300	160	2
	65	3.0	500	100	4,5
	95	3.9	300	50	8
Mi-2 B/GTD 350	15	2.8	1100	250	15
	65	3.0	700	50	32
	95	3.9	200	5	60
Mi-17/TW3-117	15	2.4	250	100	32
	65	2.5	200	50	45
	95	2.9	100	40	60

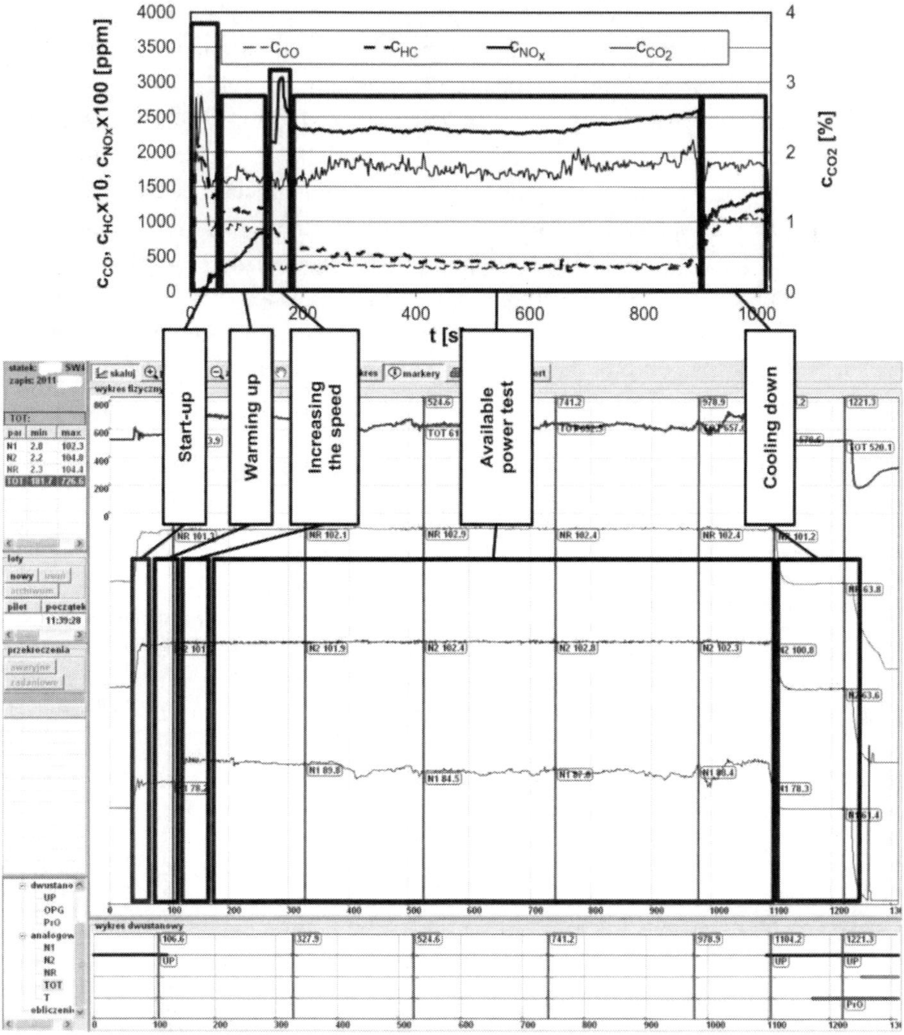

Figure 3.19. Comparison of changes in toxic compounds' concentrations against changes in engine parameters (PZL SW-4 Puszczyk).

Known engine operation parameters can be compared against available characteristics containing data on maximum load parameters generated at certain air consumption values, thus obtaining certain instantaneous fuel consumption values. On that basis, it is possible to estimate the mass of exhaust gas leaving the engine.

For the analysis in question, air consumption values declared in engine characteristics were used, and the mass of consumed fuel was determined based on measurements. The exhaust gas analyzer used in the study made it possible to measure the concentration of all toxic compounds, as well as the concentration of oxygen in exhaust gas. The excess air coefficient λ was determined on that basis.

Knowing the air consumption G_a and the excess air coefficient λ, as well as the theoretical demand for air required for full and complete combustion of 1 kilogram of fuel, fuel consumption G_e can be derived from the following formula:

$$G_e = \frac{G_p}{L_t \cdot \lambda} \text{ [kg/s]} \tag{3.1}$$

For instance, at the maximum load of the PZL-10W engine, air consumption G_a = 4.5 kg/s, theoretical demand for air required for full and complete combustion of 1 kilogram of fuel L_t = 14.2 kg air/1 kg fuel, and excess air coefficient λ = 5, the fuel flow rate is:

$$G_e = \frac{4.5 \ \frac{\text{kg}}{\text{s}}}{14.2 \ \frac{\text{kg air}}{\text{kg fuel}} \cdot 5} = 0.0632 \ \frac{\text{kg}}{\text{s}} \tag{3.2}$$

If the resultant fuel consumption value G_e is added to the air consumption value G_a, the result will constitute the mass of output exhaust gas G_{exh}. For the above example:

$$G_{exh} = G_a + G_e = 4.5 \ \frac{\text{kg}}{\text{s}} + 0.063 \ \frac{\text{kg}}{\text{s}} = 4.563 \ \frac{\text{kg}}{\text{s}} \tag{3.3}$$

Once the instantaneous exhaust gas flow rate G_{exh} is known and so are the instantaneous concentrations of toxic compounds in exhaust gas, instantaneous flow rates of toxic compounds emitted by the engines of the selected helicopters can be calculated (Figures 3.20–3.24).

The curves representing the recorded concentrations of each compound in exhaust gas emitted by the engines of the tested helicopters take similar shapes. The initial start-up phase is characterized by a significant growth in hydrocarbons concentration, which is typical for turbine engines and closely related to the ignition initiation mode in the combustor. Once the supplied fuel is ignited, carbon monoxide concentration grows rapidly. After the engine has started, hydrocarbons and carbon monoxide concentrations go down quickly to reach a certain steady level, near each engine's characteristic value. Reaching that value indicates that the engine's thermal state has stabilized. As the engine warms up, the concentration of nitrogen oxides grows. Nitrogen oxides and carbon dioxide concentrations in exhaust gas from a warm engine change depending on engine loads. Therefore, it can be claimed that information on engine loads can be inferred from records showing the concentration of these two compounds. Emission rates of all compounds depend on their respective concentrations and on exhaust gas mass. The mass of exhaust gas generated by turbine engines depends on the mass of air and fuel supplied to the engine. The mass of air depends on the compressor speed, while the amount of fuel depends on fuel supply system settings. An increase in the mass of fuel supplied to the combustor generates a greater mass of exhaust gas, which increases the power of the turbine that drives the compressor and causes an increase in its speed, thus contributing to a greater amount of air being fed to the combustor. The mass of exhaust gas depends on the demand for exhaust gas directed to the power turbine in order to generate a given level of power at the engine's propeller shaft. Therefore, emission rates of all toxic compounds in exhaust gas are linked both to their respective concentrations and to engine loads.

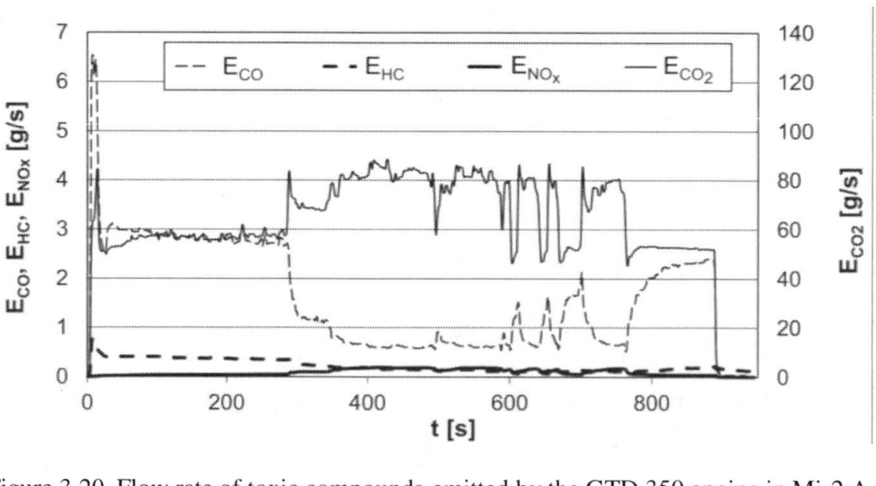

Figure 3.20. Flow rate of toxic compounds emitted by the GTD 350 engine in Mi-2 A.

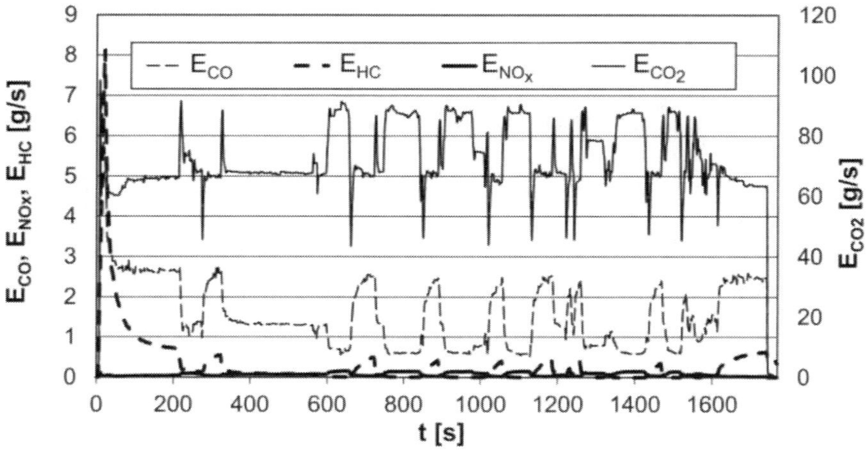

Figure 3.21. Flow rate of toxic compounds emitted by the GTD 350 engine in Mi-2 B.

Table 3.10. Emission rates of CO_2, CO, HC, NO_x for selected engine loads of the analyzed helicopters during pre flight tests

Helicopter/engine	t [s]	$N_e/N_{e\,max}$ [%]	E_{CO_2} [g/s]	E_{CO} [g/s]	E_{HC} [g/s]	E_{NOx} [g/s]
Mi-2 A/ GTD 350	28	15	49.8	2.70	0.399	0.019
	359	65	76.2	0.80	0.189	0.188
	423	95	84.5	0.60	0.155	0.193
Mi-2 B/ GTD 350	52	15	61.2	2.72	1.110	0.040
	1008	65	74.48	0.78	0.436	0.096
	1412	95	87.6	0.58	0.075	0.142
Mi-17/ TW3-117	60	15	307	3.69	2.01	0.423
	285	65	165	1.89	0.67	0.223
	736	95	352	1.72	0.53	0.753
PZL W-3 Sokół/ PZL-10W	130	15	88.1	1.41	0.0610	0.164
	165	65	63.8	0.54	0.1083	0.133
	278	72	120	0.88	0.0041	0.270
	440	84	130	0.59	0.0013	0.299
PZL SW-4 Puszczyk/ RR250-C20R/2	44	15	84.7	5.01	0.72	0.012
	132	65	83.0	4.72	0.63	0.041
	550	95	92.4	1.68	0.15	0.157

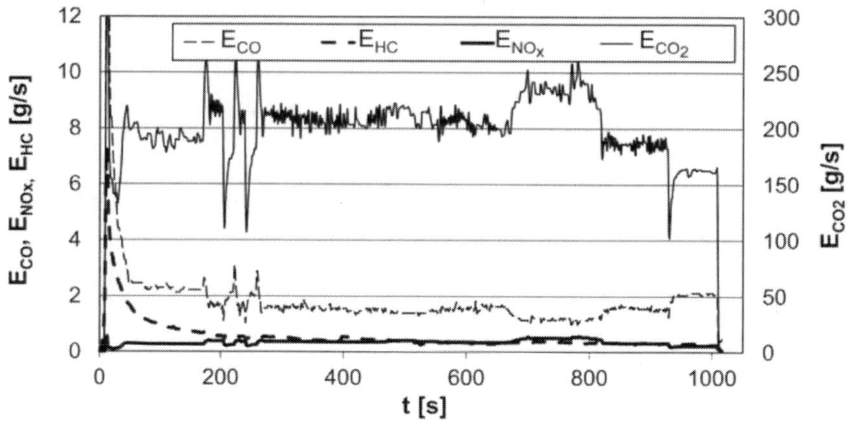

Figure 3.22. Flow rate of toxic compounds emitted by the TW3-117 engine in Mi-17.

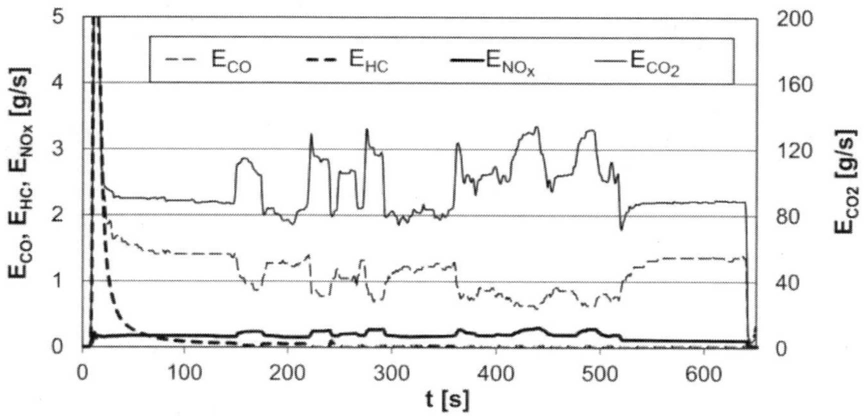

Figure 3.23. Flow rate of toxic compounds emitted by the PZL-10W engine in PZL W-3 Sokół.

On the basis of a comparison of changes in toxic compounds' concentrations against changes in operation parameters recorded during pre-flight tests for respective engine loads, emission rate read-out points were determined for the analyzed helicopter engines (Table 3.10).

An analysis of changes in emission rates of each compound and corresponding time shares of given loads allows one to calculate the mass of toxic compounds emitted during pre-flight tests of the selected helicopters (Figures 3.25–3.29).

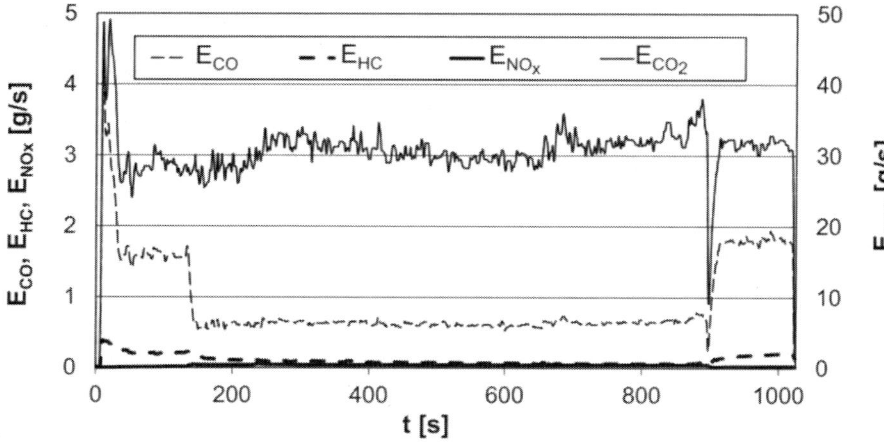

Figure 3.24. Flow rate of toxic compounds emitted by the RR250-C20R/2 engine in PZL SW-4 Puszczyk.

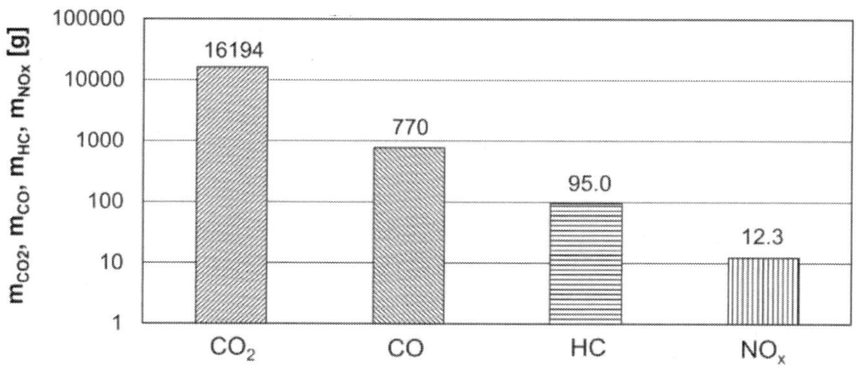

Figure 3.25. Mass of toxic compounds in exhaust gas generated by one engine of Mi-2 A during the pre-flight test (test duration: 945 s).

Each compound's mass constitutes an individual value for the analyzed engine and largely depends on test duration and engine load values during the test. The above values cannot be used for comparative analyses. For comparative purposes and in order to assess emissions from individual helicopters it is necessary to refer to specific emissions.

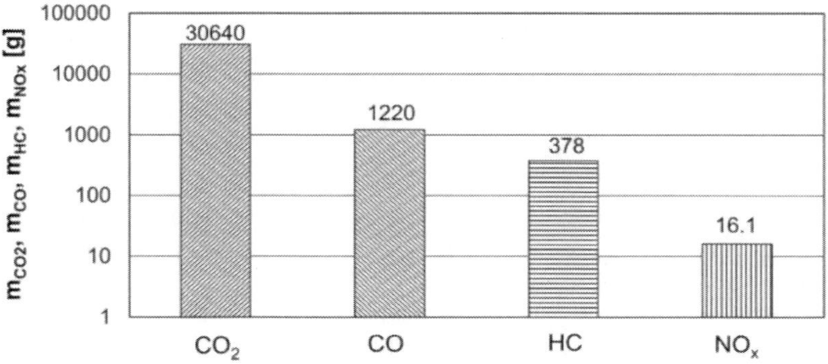

Figure 3.26. Mass of toxic compounds in exhaust gas generated by one engine of Mi-2 B during the pre-flight test (test duration: 1770 s).

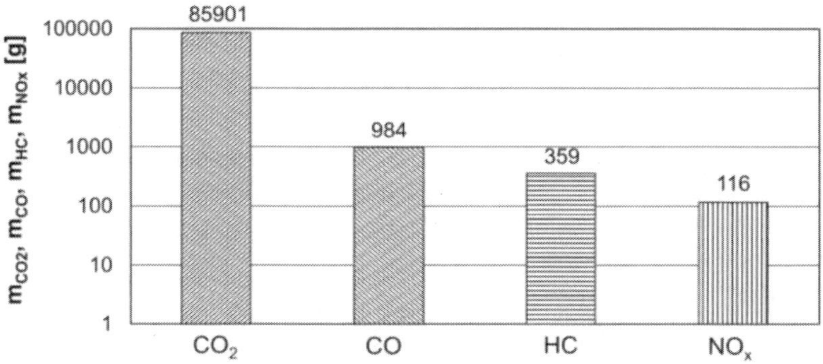

Figure 3.27. Mass of toxic compounds in exhaust gas generated by one engine of Mi-17 during the pre-flight test (test duration: 1016 s).

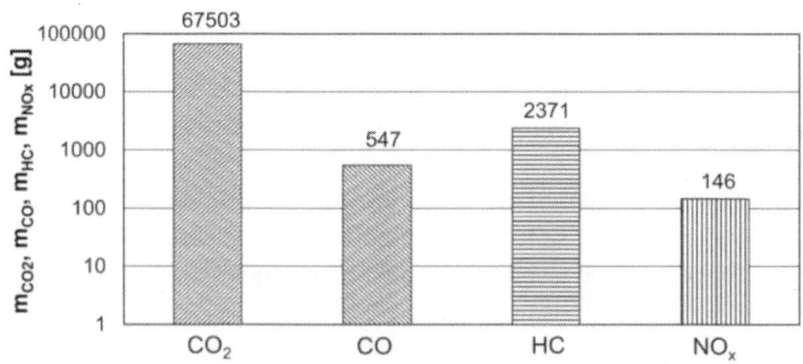

Figure 3.28. Mass of toxic compounds in exhaust gas generated by one engine of PZL W-3 Sokół during the pre-flight test (test duration: 650 s).

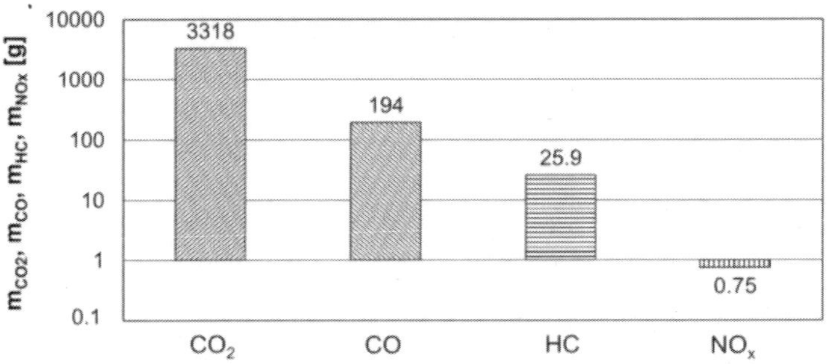

Figure 3.29. Mass of toxic compounds in exhaust gas generated by the engine of PZL SW-4 Puszczyk during the pre-flight test (test duration: 1024 s).

The aforementioned emission rates of each toxic compound were compared against engine loads calculated for Mi-2 and Mi-17, and those recorded by on-board recorders for PZL W-3 Sokół and PZL SW-4 Puszczyk. The said comparison made it possible to determine specific emissions of toxic compounds in exhaust gas for different engine load points (Figures 3.30–3.34).

Available data show that the growth in engine load is accompanied by reduced specific emissions of toxic compounds in exhaust gas. The said tendency is particularly visible in the case of carbon monoxide and hydrocarbons. Quite importantly, low engine load is characteristic of the shortest operation phase, i.e., starting up and warming up the engine. For a warm engine, the concentration of these compounds is many times lower. For a warm engine, specific emissions of carbon dioxide and nitrogen oxides decrease only slightly in the load range of 65–84% $N_e/N_{e\,max}$.

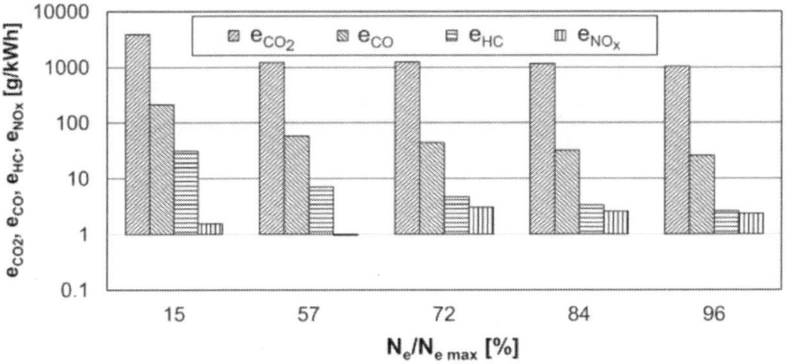

Figure 3.30. Specific emission of toxic compounds in exhaust gas generated by the propulsion system of Mi-2 A for various load values.

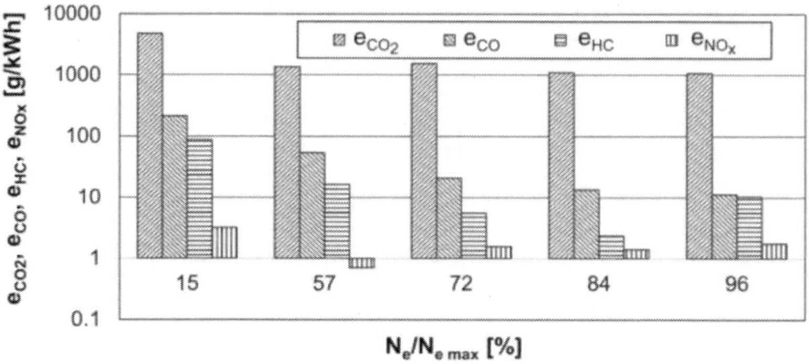

Figure 3.31. Specific emission of toxic compounds in exhaust gas generated by the propulsion system of Mi-2 B for various load values.

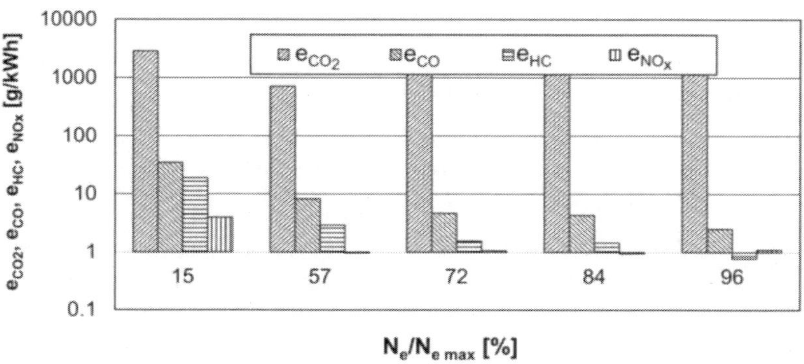

Figure 3.32. Specific emission of toxic compounds in exhaust gas generated by the propulsion system of Mi-17 for various load values.

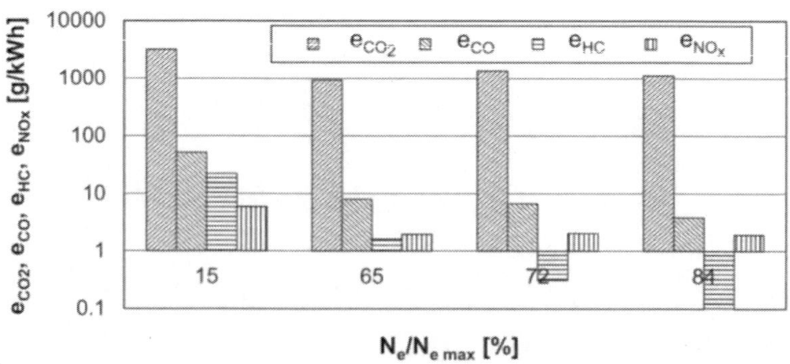

Figure 3.33. Specific emission of toxic compounds in exhaust gas generated by the propulsion system of PZL W-3 Sokół for various load values.

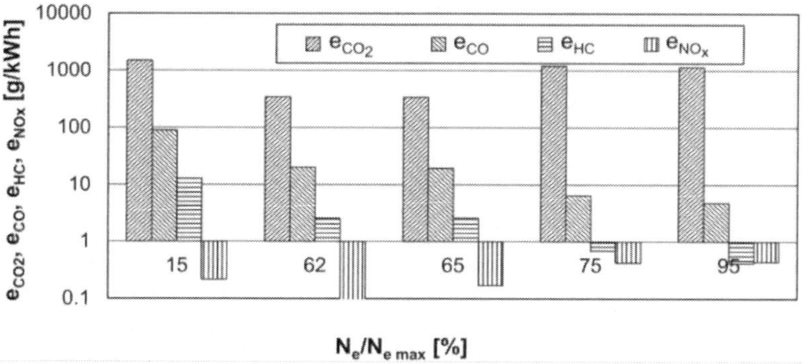

Figure 3.34. Specific emission of toxic compounds in exhaust gas generated by the propulsion system of PZL SW-4 Puszczyk for various load values.

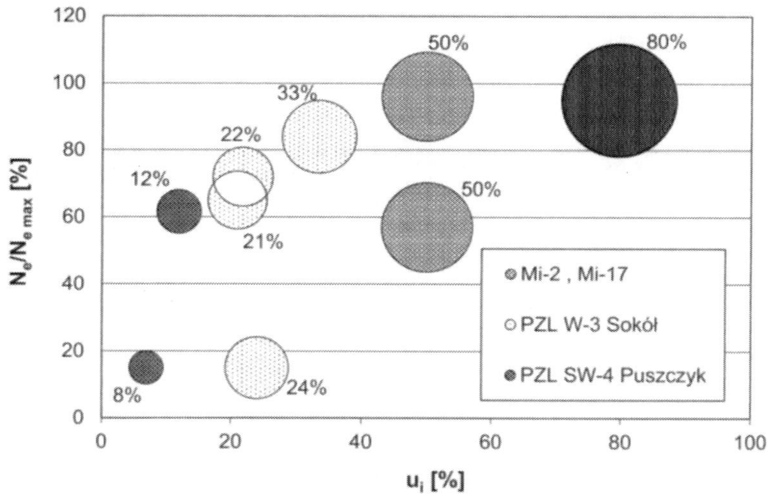

Figure 3.35. Time loads of helicopter propulsion systems during the pre-flight test (circle size denotes relative time share).

Specific emissions of toxic compounds in exhaust gas generated by propulsion systems of the selected helicopters, as observed in operation conditions during pre-flight tests, depend on engine load values. Test scenarios were not identical for all tested helicopters, but the relative share of engine load time and respective load percentages are similar to some extent (Figure 3.35). A comparison of engine load times recorded during the tests as well as test guidelines for Mi-2 and Mi-17 allows one to determine the mean engine load during the pre-flight test, calculated as follows:

$$\left(\frac{N_e}{N_{e\,max}}\right)_{avg} = \sum \frac{N_e}{N_{e\,max}} \cdot u_i \qquad (3.4)$$

where:

$N_e/N_{e\,max}$ instantaneous relative engine load value [%],

u_i share in overall test time [–].

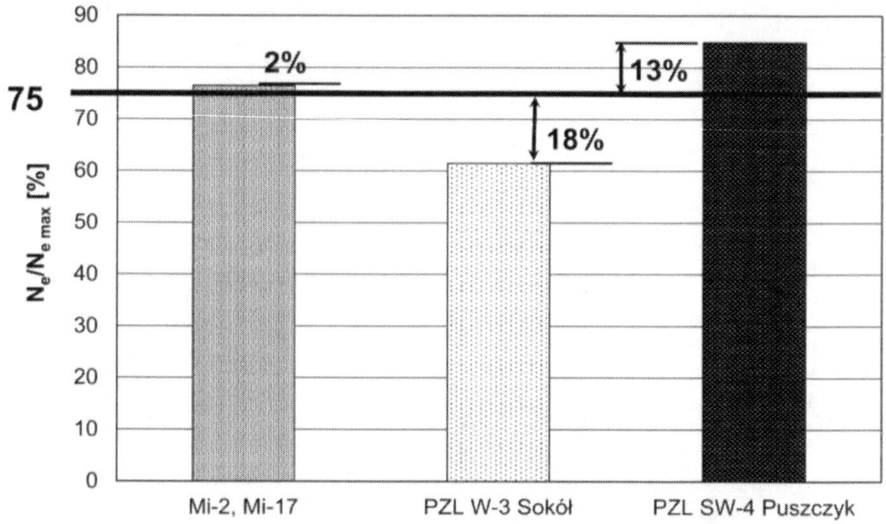

Figure 3.36. Mean helicopter engine loads during pre-flight tests.

To calculate mean load values for all tested helicopters, individual $N_e/N_{e\,avg}$ values were aggregated and then divided by the number of tested aircraft. As a result, the mean engine load for all helicopters in all pre-flight tests was calculated and amounts to 75% of the maximum load (Figure 3.36). It needs to be remarked that mean pre-flight engine loads for each helicopter do not differ from the grand mean by more than 20%. The difference of 18% observed for PZL W-3 Sokół and 13% for PZL SW-4 Puszczyk is attributable to extended test duration.

To determine the values of specific toxic compound emissions characteristic of a given engine in a given pre-flight test, specific emissions of toxic compounds in exhaust gas determined for individual load ranges should be multiplied by the relative time share of those ranges in the overall pre-flight test time:

$$e_j = \sum e_{ji} \cdot u_i \left[\frac{\text{g}}{\text{kWh}} \right] \quad (3.5)$$

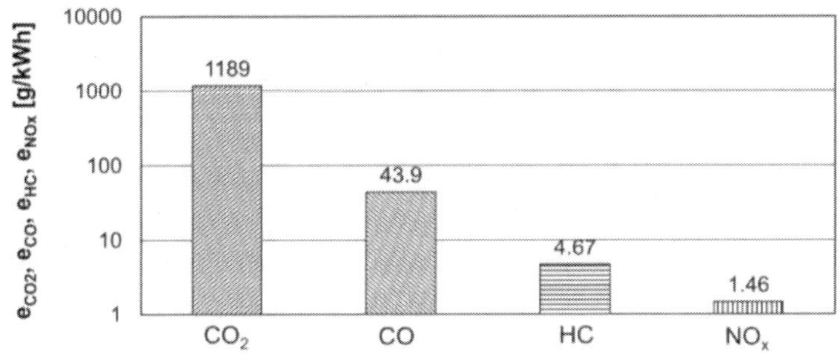

Figure 3.37. Specific emission of toxic compounds in exhaust gas generated by the propulsion system of Mi-2 A during the pre-flight test.

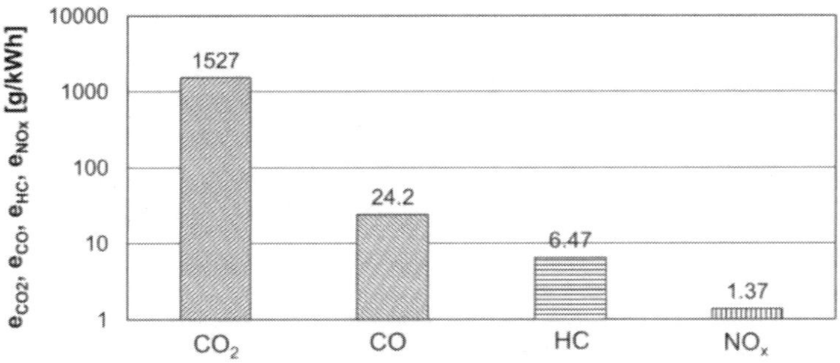

Figure 3.38. Specific emission of toxic compounds in exhaust gas generated by the propulsion system of Mi-2 B during the pre-flight test.

where:

e_{ji} specific emission for a given load [g/kWh],
u_i relative share in the overall test time [–].

Mean specific emission values determined for pre-flight tests of each helicopter are presented in the charts below (Figures 3.37–3.41).

For the Mi-2 A's engines, specific emissions of toxic compounds in exhaust gas during the pre-flight test were as follows: e_{CO2} = 1189 g/kWh, e_{CO} = 43.9 g/kWh, e_{HC} = 4.67 g/kWh, e_{NOx} = 1.46 g/kWh. For the Mi-2 B's engines, the corresponding values were: e_{CO2} = 1527 g/kWh, e_{CO} = 24.2 g/kWh, e_{HC} = 6.47 g/kWh, e_{NOx} = 1.37 g/kWh. For the engines of Mi-17, specific emissions of toxic compounds in exhaust gas during the pre-flight test were as follows: e_{CO2} = 1252 g/kWh, e_{CO} = 15.7 g/kWh, e_{HC} = 1.99 g/kWh, e_{NOx} = 1.85 g/kWh. For the engines of PZL W-3 Sokół: e_{CO2} = 1317 g/kWh, e_{CO} = 16.0 g/kWh, e_{HC} = 2.81 g/kWh, e_{NOx} = 2.76 g/kWh. For the engines of PZL SW-4 Puszczyk specific emissions of toxic compounds in exhaust gas during the pre-flight test were as follows: e_{CO2} = 1456 g/kWh, e_{CO} = 19.79 g/kWh, e_{HC} = 2.99 g/kWh, e_{NOx} = 1.17 g/kWh.

Specific emissions of toxic compounds in exhaust gas generated by propulsion systems of each helicopter show a somewhat individualized character of every engine, particularly visible in the case of carbon monoxide and hydrocarbons. Specific emissions of these compounds suggest a possibility of introducing a division based on the design of the analyzed engines. For engines with an annular combustor, specific emission of carbon monoxide does not exceed 16 g/kWh, and specific emission of hydrocarbons does not exceed 2.9 g/kWh. For engines with a can combustor, specific emission of carbon monoxide exceeds 19 g/kWh, while specific emission of hydrocarbons exceeds 2.9 g/kWh. As regards specific emission of nitrogen oxides, it does not exceed 1.5g/kWh in can combustor engines. For annular combustor engines, specific emission of nitrogen oxides is above 1.5 g/kWh. Identification of characteristic features of engine designs and their effect on specific emissions of individual toxic compounds is a fundamental step towards engine diagnostic assessments based on toxic emissions in exhaust gas.

On the basis of emission rates of toxic compounds in exhaust gas from helicopter engines on the one hand and in-flight engine parameters on the other one can assess – in a similar manner – emissions from helicopters in actual operation conditions.

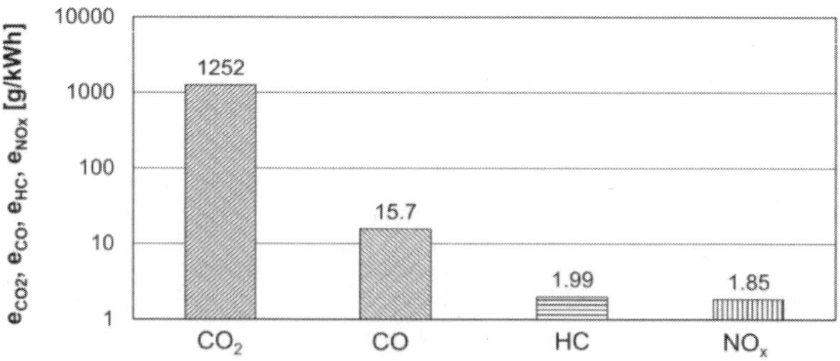

Figure 3.39. Specific emission of toxic compounds in exhaust gas generated by the propulsion system of Mi-17 during the pre-flight test.

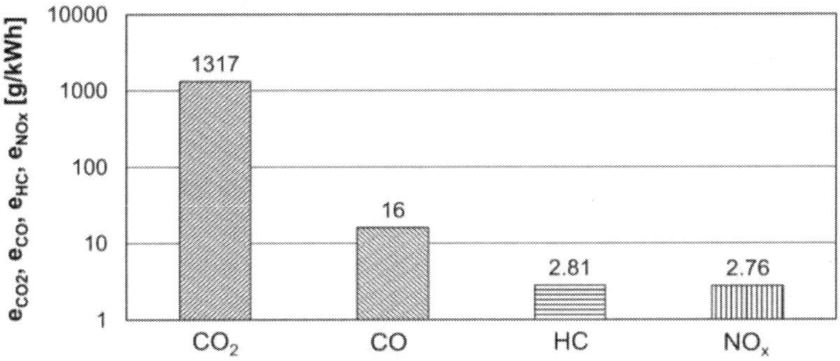

Figure 3.40. Specific emission of toxic compounds in exhaust gas generated by the propulsion system of PZL W-3 Sokół during the pre-flight test.

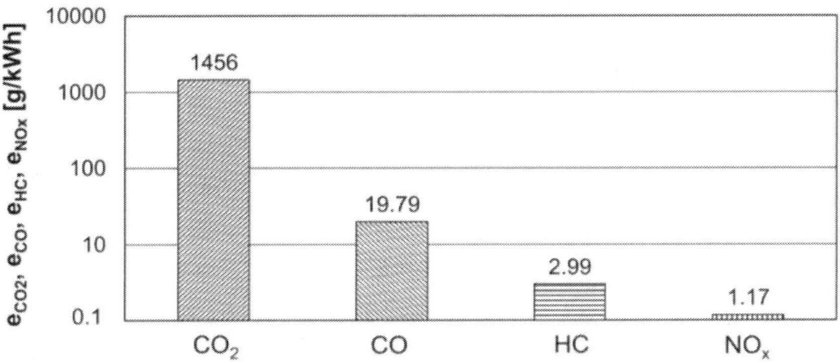

Figure 3.41. Specific emission of toxic compounds in exhaust gas generated by the propulsion system of PZL SW-4 Puszczyk during the pre-flight test.

3.5. Emissions from Helicopters in Actual Operation Conditions

An analysis of helicopter engine loads and their respective durations in actual operation conditions is possible for aircraft fitted with on-board flight parameter recorders. On-board recorders make it possible to record selected parameters as a function of time. For the purpose of this study, several sets of parameters recorded by on-board recorders of the selected helicopters were analyzed. Available records came from approximately fifteen different flights. Out of all available sets, 10 were randomly selected and analyzed (Figure 3.42).

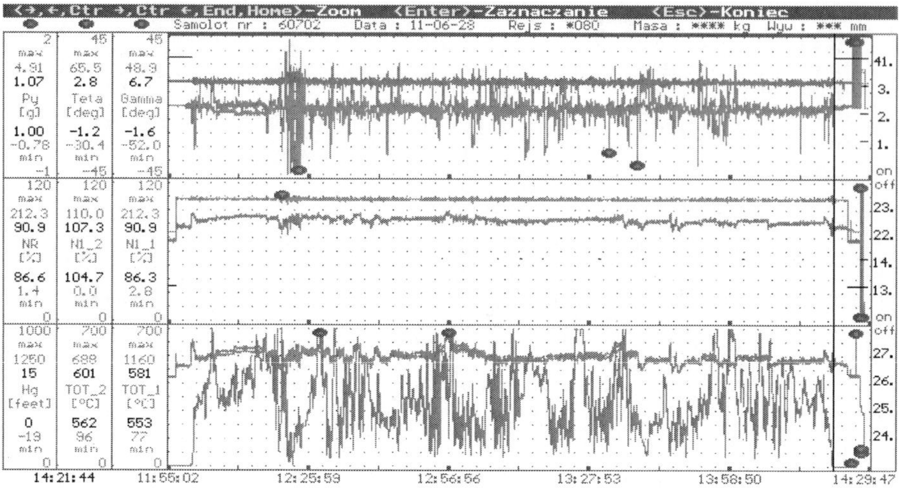

Figure 3.42. Sample record of engine operation parameters during a flight of W-3 Sokół (source: on-board flight recorder).

A time analysis of the PZL W-3 Sokół's engine loads showed that for approximately 96% of the time the engines operate at 80–90% of the maximum load (Figure 3.43).

The parameters recorded during 10 randomly selected flights (Figure 3.44) in actual helicopter operation were analyzed as well. Engine load histograms (Figure 3.45) showed that for approximately 95% of the time the engine operates at approximately 95% of the maximum load, while partial loads (approx. 60%) correspond to approximately 4% of total operation time.

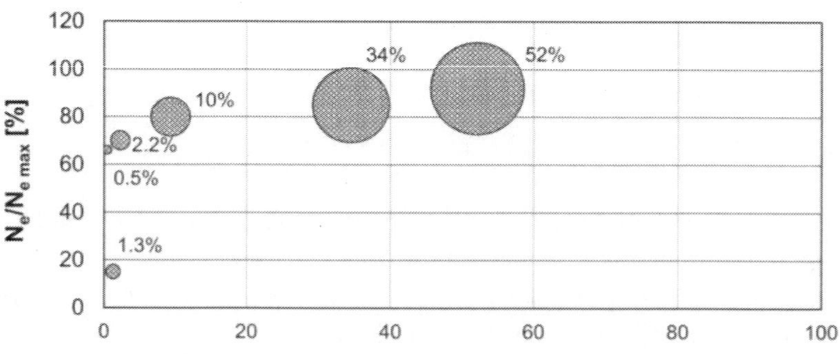

Figure 3.43. A histogram of the PZL W-3 Sokół's propulsion system loads in actual operation conditions (circle size denotes relative time share).

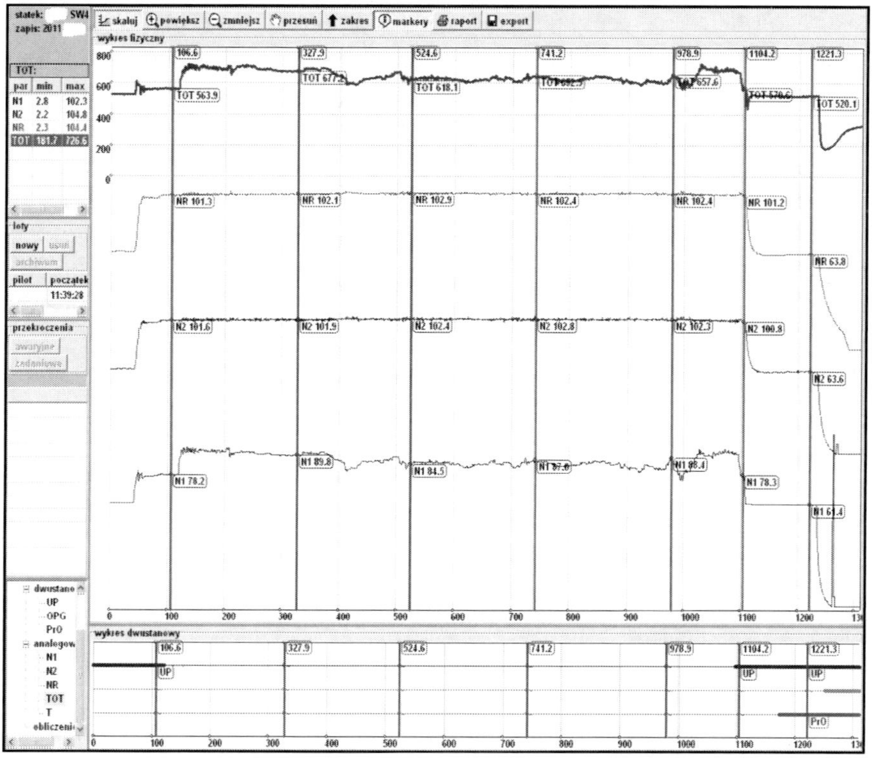

Figure 3.44. Sample record of engine operation parameters during a flight of SW-4 Puszczyk (source: on-board flight recorder).

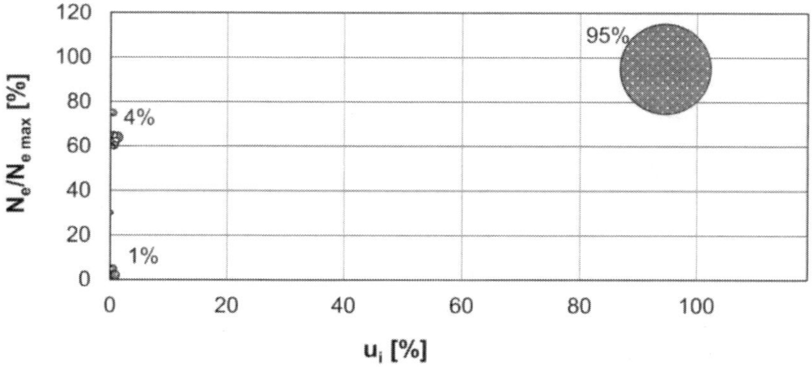

Figure 3.45. A histogram of the SW-4 Puszczyk's propulsion system loads in actual operation conditions (circle size denotes relative time share).

Data on helicopter engine operation conditions obtained based on flight recorder charts were compared against instantaneous concentrations of toxic compounds in exhaust gas. As a result, emission rates of those compounds corresponding to a given engine load were calculated. The observed dependencies were subsequently analyzed in order to assess emissions during helicopter flights.

The above relative shares of engine load times in the total flight time were multiplied by specific emission values corresponding to given engine loads as per equation (3.5). The products of these multiplications were specific emissions of toxic compounds in exhaust gas generated by the propulsion system during flights of PZL W-3 Sokół (Figure 3.46) and PZL SW-4 Puszczyk (Figure 3.47).

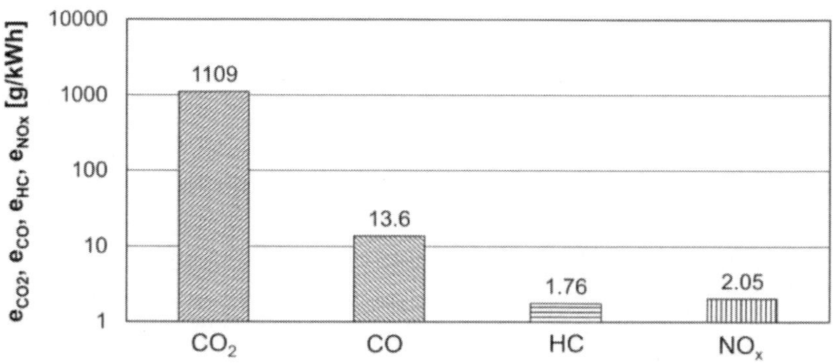

Figure 3.46. Specific emission of toxic compounds in exhaust gas from PZL W-3 Sokół in actual operation conditions.

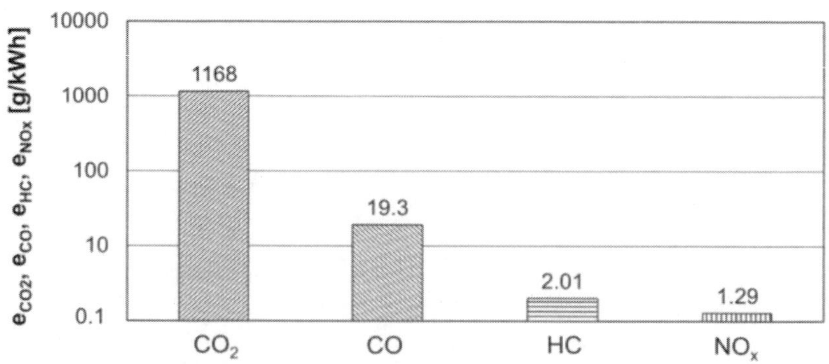

Figure 3.47. Specific emission of toxic compounds in exhaust gas from PZL SW-4 Puszczyk in actual operation conditions.

For the PZL-10W engine powering PZL W-3 Sokół, specific emissions of toxic compounds in exhaust gas were as follows: 1109 g/kWh for carbon dioxide; 13.6 g/kWh for carbon monoxide; 1.76 g/kWh for hydrocarbons; 2.05 g/kWh for nitrogen oxides. For the Rolls-Royce 250-C20R/2 engine powering PZL SW-4 Puszczyk, specific emissions of toxic compounds in exhaust gas during the pre-flight test were as follows: 1168 g/kWh for carbon dioxide; 19.3 g/kWh for carbon monoxide; 2.01 g/kWh for hydrocarbons and 1.29 g/kWh for nitrogen oxides.

Significant differences can be observed as regards specific emissions of toxic compounds in exhaust gas generated by the engines of PZL W-3 Sokół on the one hand and PZL SW-4 Puszczyk on the other. Specific carbon dioxide emission from the PZL-10W engines is similar to those from Rolls Royce 250-C20R/2. Specific nitrogen oxides emission from the PZL-10W engine is by approximately 80% greater than from the other engine. However, specific emissions of carbon monoxide from the PZL-10W engine is by approximately 70% lower than from the Rolls Royce 250-C20R/2 engine. For hydrocarbons the relative difference is approximately 20%, to the benefit of the PZL-10W engine. The differences can be explained by the designs of both engines. The PZL-10W engine has an annular combustor, while the Rolls Royce 250-C20R/2 engine has a can combustor. The difference in engine designs may be a reason for changes in emissions from the selected helicopters.

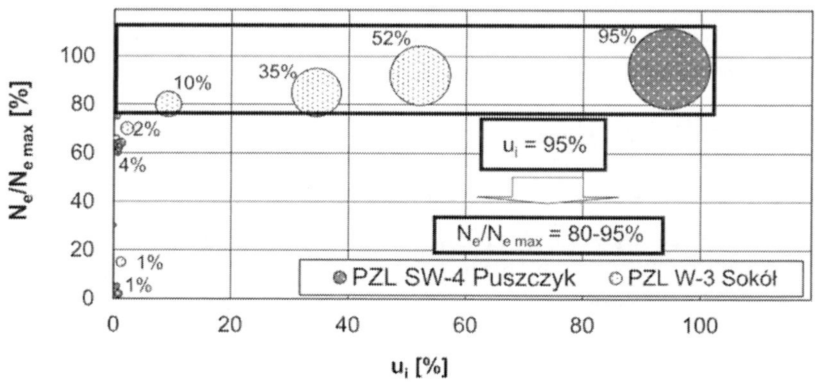

Figure 3.48. Comparison of engine load histograms for PZL W-3 Sokół and PZL SW-4 in actual operation conditions.

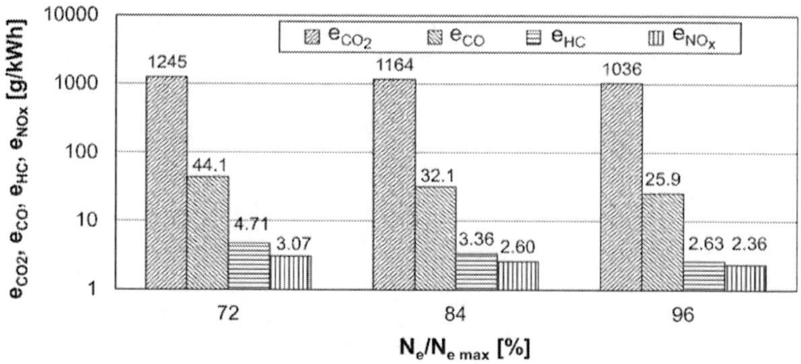

Figure 3.49. Specific emission of toxic compounds in exhaust gas generated by the propulsion system of Mi-2 A for various load values.

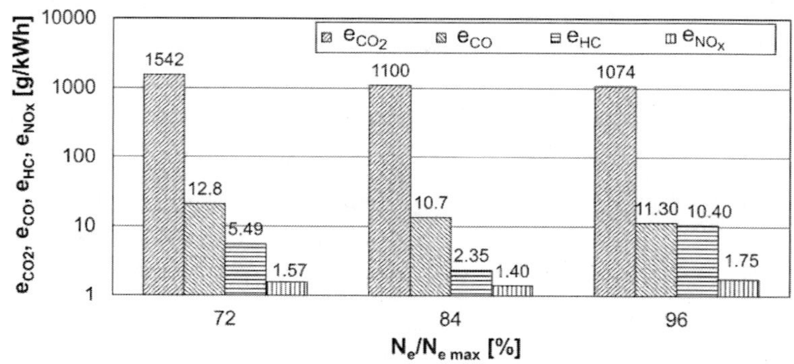

Figure 3.50. Specific emission of toxic compounds in exhaust gas generated by the propulsion system of Mi-2 B for various load values.

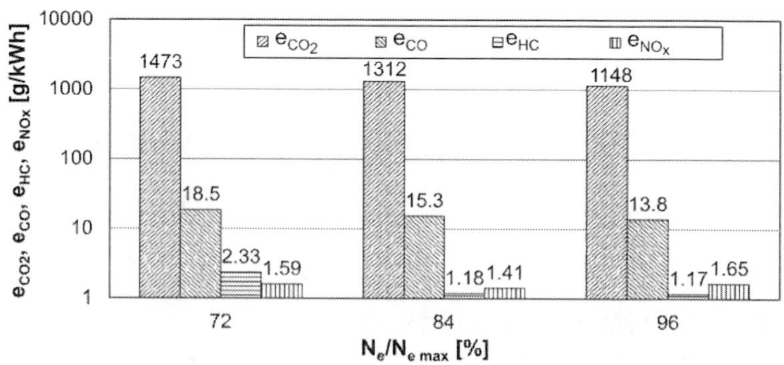

Figure 3.51. Specific emission of toxic compounds in exhaust gas generated by the propulsion system of Mi-17 for various load values.

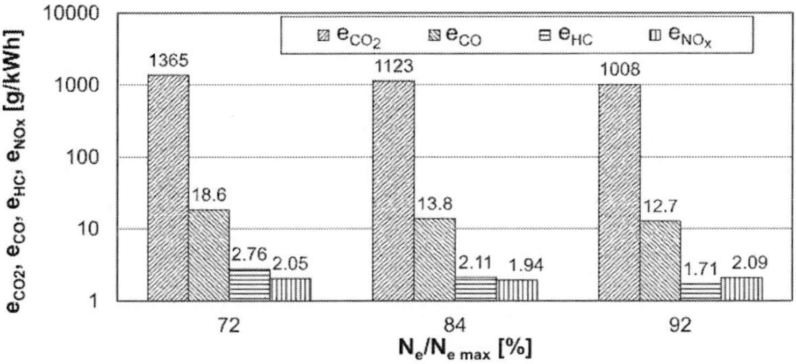

Figure 3.52. Specific emission of toxic compounds in exhaust gas generated by the propulsion system of PZL W-3 Sokół for various load values.

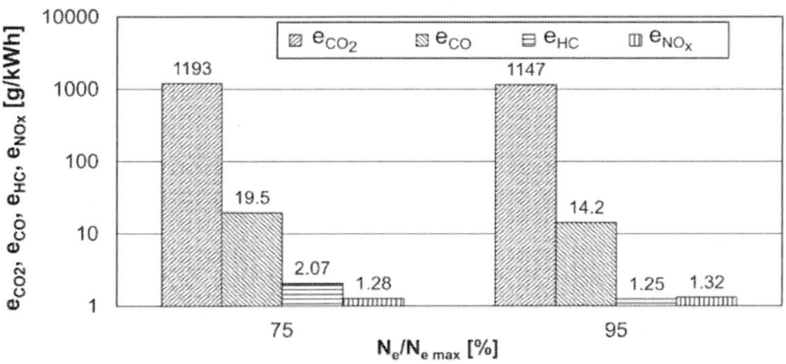

Figure 3.53. Specific emission of toxic compounds in exhaust gas generated by the propulsion system of PZL SW-4 Puszczyk for various load values.

Sets of data on propulsion system loads recorded in actual helicopter operation conditions can be compared to one another (Figure 3.48). The comparison shows that helicopter engine loads stay in the range of 80–95% of the maximum load for 95% of operation time.

The results of the analysis show that the most significant environmental impacts are caused in those operation phases where the propulsion system works at greatest loads. Toxic compound concentration measurements carried out during pre-flight tests as well as engine operation parameter records, followed by analyses of those data made it possible to assign specific emissions of each compound to their corresponding engine load states. Since the engine load varies from 80% to 95% of the maximum load, a comparison of specific emissions of toxic compounds in exhaust gas characteristic of engine operation points corresponding to the said range was carried out (Figs. 3.49–3.53). The results of a comparative analysis of selected specific emission values against specific emission at 96% of the maximum engine load are presented in Table 3.11.

The above data confirm the character of changes in specific emissions of toxic compounds caused by engine load increases. They confirm the relationships between engine load on the one hand and combustion process and its inherent reactions on the other. Toxic compounds in exhaust gas are formed during the said reactions. The above-presented percentage differences in specific emissions of toxic compounds as compared to the specific emissions value at 96% of the maximum engine load are significant and differ depending on the combustor type. The foregoing is particularly visible in the case of specific emissions of carbon monoxide and hydrocarbons. Relative differences between the compared values for two different engines (PZL-10W and TW3-117) exceed 60%. As regards engines with can combustors the differences in the emissions of those two compounds do not exceed 50%. Therefore, in order to determine specific emissions of toxic compounds in actual engine operation conditions it is necessary to carry out an analysis based on emission values for all identified operation points, taking into account their respective shares in the total operation time.

Table 3.11. Results of a comparative analysis of specific emission values for selected engine load values

Helicopter/engine	$N_e/N_{e\ max}$ [%]	Relative difference in specific emissions [%] (reference $N_e/N_{e\ max}$ = 96%)			
		Δe_{CO_2}	Δe_{CO}	Δe_{HC}	Δe_{NO_x}
Mi-2 A/ GTD 350	84	12.4	23.9	27.4	10.3
Mi-2/ GTD 350	84	2.4	19.4	125.8	–19.6
Mi-17/ TW3-117	84	14.3	10.8	86.5	–14.2
PZL W-3 Sokół/ PZL-10W	84	11.4	8.9	23.7	–6.9
PZL SW-4 Puszczyk/ RR250-C20R/2	75	4.0	37.6	65.2	–3.5

For PZL W-3 Sokół and PZL SW-4 Puszczyk it was possible to determine the operation parameters based on values recorded by their on-board flight recorders during actual operation of the helicopters. In the case of the other tested helicopters (Mi-2 and Mi-17) engine operation parameters during helicopter operation were not recorded, as those helicopters are not fitted with on-board flight recorders. In the absence of any source material for the analysis of engine load values and times during the flight it was assumed that operation conditions of these helicopters are identical to those of the PZL W-3 Sokół, as shown in Figure 3.48.

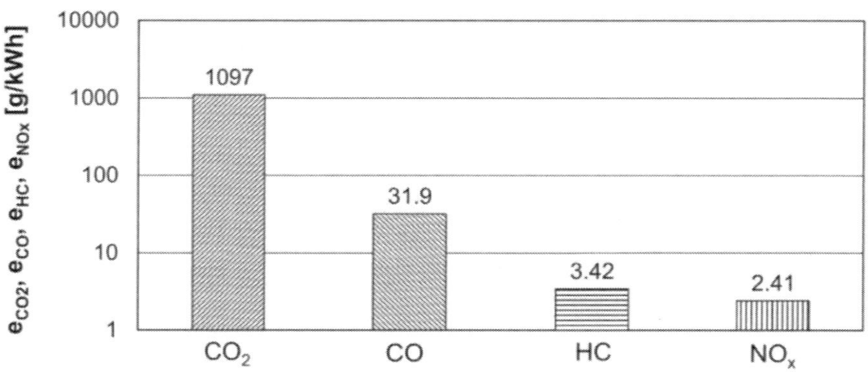

Figure 3.54. Specific emission of toxic compounds in exhaust gas generated by the propulsion system of Mi-2 A in actual operation conditions.

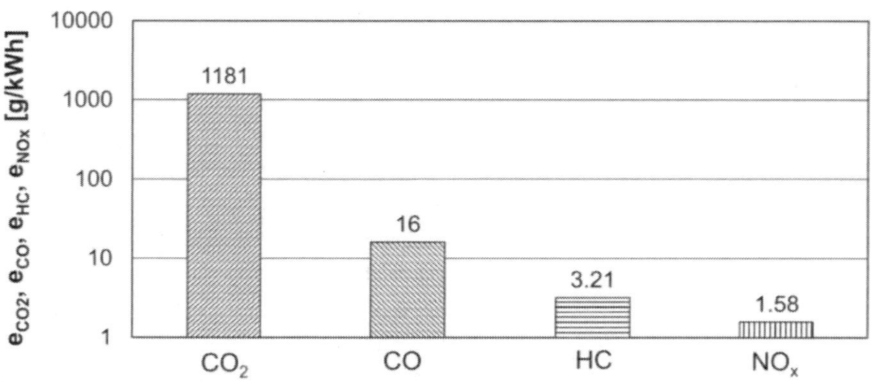

Figure 3.55. Specific emission of toxic compounds in exhaust gas generated by the propulsion system of Mi-2 B in actual operation conditions.

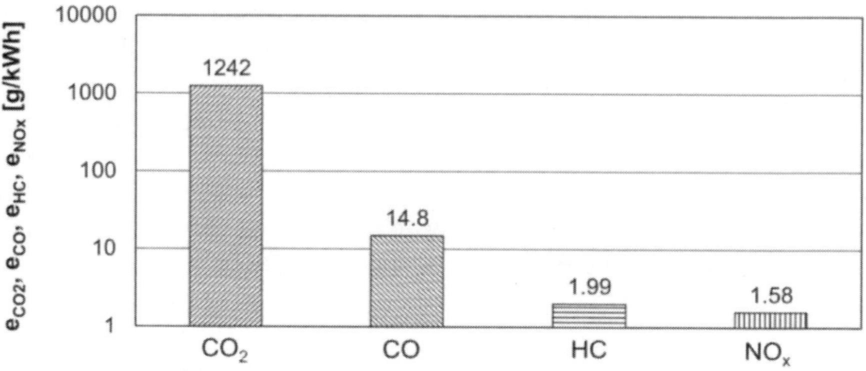

Figure 3.56. Specific emission of toxic compounds in exhaust gas generated by the propulsion system of Mi-17 in actual operation conditions.

In accordance with the procedure of the analysis presented earlier in this chapter, the assumed percentage shares of engine load times during a flight of PZL W-3 Sokół were multiplied by specific emissions corresponding to engine loads determined for Mi-2 A, Mi-2 B and Mi-17, as per equation (3.5). As a result, specific emissions of toxic compounds in exhaust gas generated by the propulsion systems of the said helicopters were calculated (Figures 3.54–3.56).

The values of toxic emissions in exhaust gas generated by the GTD-350 engine used to propel Mi-2 A are as follows: 1097 g/kWh for carbon dioxide; 31.9 g/kWh for carbon monoxide; 3.42 g/kWh for hydrocarbons; and 2.41 g/kWh for nitrogen oxides. For the GTD-350 engine used to propel Mi-2 B the values of toxic emissions in exhaust gas are as follows: 1181 g/kWh for carbon dioxide; 16.0 g/kWh for carbon monoxide; 3.21 g/kWh for hydrocarbons; and 1.58 g/kWh for nitrogen oxides. For the TW3-117 engine used to propel Mi-17 the values of toxic emissions in exhaust gas are as follows: 1242 g/kWh for carbon dioxide; 14.8 g/kWh for carbon monoxide; 1.99 g/kWh for hydrocarbons; and 1.58 g/kWh for nitrogen oxides.

In the case of specific emissions of toxic compounds in exhaust gas generated by the engines of Mi-2 and Mi-17 the following differences can be observed. Specific emission of carbon dioxide from the GTD-350 engines is by approximately 10% lower than from the TW3-117 engine. As regards specific emissions of nitrogen oxides from Mi-17 and Mi-2 B, the obtained values are identical. However, it was observed that specific emission of nitrogen oxides for Mi-2 A is by approximately 65% greater as compared to

Mi-17 and Mi-2 B. A similar situation is observed for specific emission of carbon monoxide which in Mi-2 A is twice as high as in Mi-17 and Mi-2 B. Specific emissions of hydrocarbons from Mi-2 A and Mi-2 B are by approximately 40% greater than specific emissions of hydrocarbons from Mi-17.

By comparing specific emissions of each toxic compound in exhaust gas generated by two different engines of the same design one can directly compare the environmental performance of those engines. By comparing specific emission against reference engine emission one could verify the settings of the fuel supply system and the engine's technical condition. The two Mi-2s analyzed during the study had different service lives. The engines of Mi-2 A had just undergone a major repair, while the service life of those of Mi-2 B was about to end. The comparison of specific emissions of carbon monoxide shows that their values do differ: for the GTD-350 engine fitted in Mi-2 A they are approximately twice as high as those for the GTD-350 engine fitted in Mi-2 B. The underlying reason could be possible misalignment of the fuel supply system or the engine's poor technical condition. In the case of these two analyzed helicopters, the obtained results explain the external appearance of the helicopters. Despite the engines' recent major repair, Mi-2 A generates a lot of smoke during operation, which is testified by dirt accumulated on the fuselage from the exhaust gas outlet over to the tail rotor.

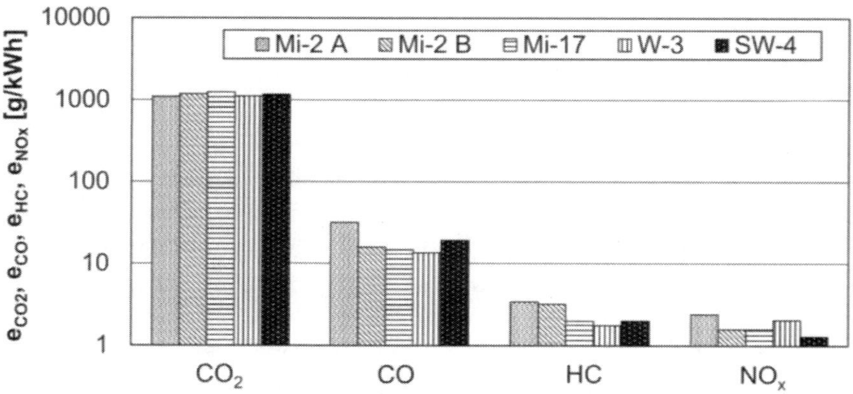

Figure 3.57. Specific emission of toxic compounds in exhaust gas from all study helicopters in actual operation conditions.

To illustrate the issue of excessive specific emissions of toxic compounds generated by the engines of Mi-2 A, a comprehensive comparison of the obtained emission values was carried out (Figure 3.57). The comparison shows that the greatest specific emission of toxic compounds in exhaust gas from turbine engines used to propel the selected helicopters is generated by the GTD-350 engines fitted in Mi-2s. These engines have a can combustor and are affected by misalignment of the fuel supply system (Mi-2 A) or an excessively long engine service time (Mi-2 B).

Helicopter tests followed by an analysis of measured concentrations of emissions (both in pre-flight conditions and in actual operation conditions) indicate that it is possible to assess the technical condition of helicopter engines by means of measuring the concentration of selected compounds in exhaust gas. To ensure unequivocal evaluation of environmental performance of the engine it is necessary to develop a procedure for a helicopter engine emission test.

3.6. Exhaust Emissions Test for Helicopter Turbine Engines

Verification of toxic emissions in exhaust gas from helicopter propulsion systems involves measurements of such emissions in actual operation conditions. For formal reasons, the entities in charge of the study helicopters refused to enable emission measurements in actual operation conditions. Therefore, emissions were measured during pre-flight tests, as discussed in the preceding chapters. The measurements and the analyses of their results showed that it is possible to diagnose turbine engines used for helicopter propulsion on the basis of information on toxic compounds' concentration in exhaust gas. Therefore it would be recommendable do develop a test serving the purpose of measuring emissions of toxic compounds from helicopter turbine engines.

The awareness of helicopter emissions' evaluation criteria presented in this publication makes it possible to develop an all-purpose test for assessing emissions of toxic compounds generated by propulsion systems of transport helicopters. Development and implementation of such a test with regard to diagnostic procedures could contribute to reducing the time and workload necessary to verify the technical condition of such systems. Therefore, an attempt was taken at developing an emissions test based on the following assumptions:

- the test should be representative of all transport helicopters powered with turbine engines; exhaust gas emission results from a static test should be comparable to the results obtained in actual operation conditions;
- number of test phases should not be greater than 3, and they should be designed so that they closely reflect the most common engine load ranges observed in actual operation (idling, partial load and maximum load);
- the test should be incorporated in pre-flight test procedures.

The importance of the third of the above assumptions follows from the fact that pre-flight test duration counts towards helicopter engine service life. Therefore, it would be advisable to avoid reducing the remaining service life by running emission tests separately from pre-flight tests.

The starting point for designing the test was also based on operation parameters recorded for selected flights of PZL W-3 Sokół and PZL SW-4 Puszczyk, taken into account in the analyses described in chapter 3.5. On the basis of the above assumptions and the engine load time share characteristics, the engines' operation range was divided into three areas, separately for PZL W-3 Sokół (Figure 3.58) and PZL SW-4 Puszczyk (Figure 3.59): low load operation range (range A), medium load operation range (range B) and maximum load operation range (range C). Range A corresponds to engine start-up, during which exhaust gas generator turbine speeds stay in the range of 80–100% of the maximum load.

By comparing the relative percentage of time corresponding to each engine load range during flights and during pre-flight tests one can conclude that engine operation load ranges are similar. Therefore, it is possible to devise an emission test dedicated to toxic compounds in helicopter exhaust gas that can be carried out during pre-flight tests.

In order to develop such a test and to interpret the results of emission measurements it is necessary to determine coefficients representing the share of each test phase in the measured emission of toxic compounds.

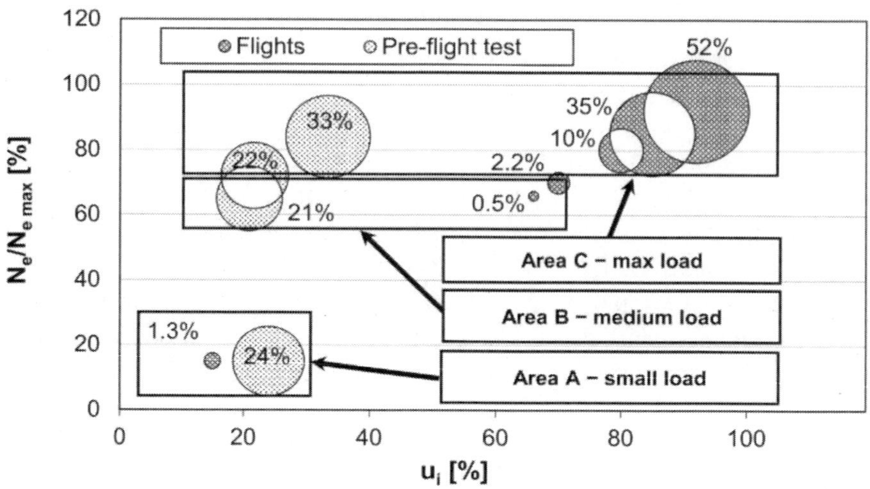

Figure 3.58. Test phase selection methodology based on engine load time shares for PZL W-3 Sokół (flights and pre-flight test).

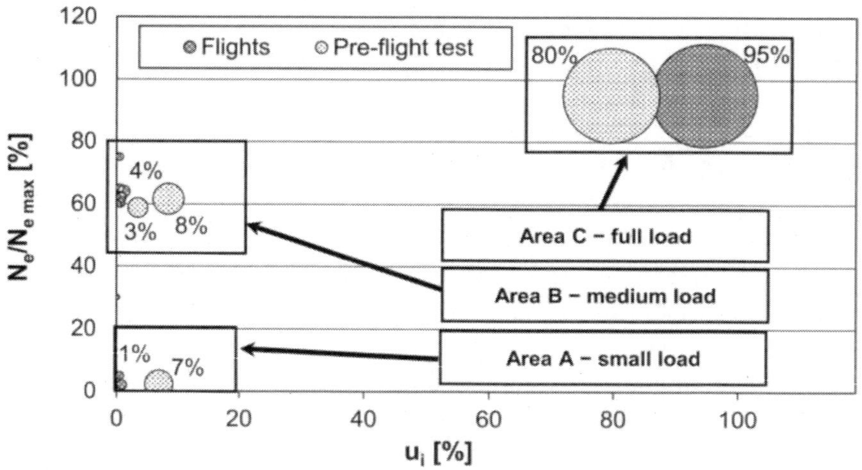

Figure 3.59. Test phase selection methodology based on engine load time shares for PZL SW-4 Puszczyk (flights and pre-flight test).

Weight coefficients were selected based on time shares of instantaneous engine load values recorded by the on-board flight recorder during ten randomly selected helicopter flights. In consistence with the analysis carried out in chapter 3.5, related to the assessment of performance parameters of the engines fitted in PZL W-3 Sokół and PZL SW-4 Puszczyk recorded in actual operation conditions, percentage-based time shares for all load ranges were introduced (Table 3.12). Pursuant to the underlying assumption for the test (suitability for all turbine engine powered helicopters), test phase shares were averaged. Figure 3.60 demonstrates the shares and the distribution of individual test phases. For such mean phase shares, absolute error values were determined (individually for PZL W-3 Sokół and PZL SW-4 Puszczyk). Absolute error values determined for mean phase shares are as follows: phase A – approx. 11%, phase B – approx. 24%, phase C – approx. 0.5%. The smallest absolute error value was determined for phase C, representing the greatest share in the overall test time. Therefore, it is likely that emission values arrived at in the test will largely reflect the emission values from actual operation conditions.

The developed stationary emission test dedicated to turbine engines used in helicopters needs to be verified. Such verification will make it possible to assess the test's relevance in determining actual emissions from transport helicopters. The following subchapter presents the test's verification results for selected aircraft.

3.7. Emission Test Verification

Emissions from the selected helicopters during emission tests were assessed on the basis of measured concentrations of toxic compounds in exhaust gas recorded during pre-flight tests. On the basis of measured concentrations and engine operation parameters recorded by on-board flight recorders in actual operation conditions, possible in-flight emissions were analyzed. The obtained specific emission values for each toxic compound corresponding to all engine operation states were used to assess the correctness of the all-purpose helicopter emission test.

Specific emissions of toxic compounds in exhaust gas expressed in g/kWh were used as reference values for test verification. The data used in the comparative analysis are shown in Table 3.13, presenting specific emissions of toxic compounds at selected operation points of the engines of PZL W-3 Sokół, PZL SW-4 Puszczyk, Mi-2 A, Mi-2 B and Mi-17. Sets of data for each of the analyzed aircraft were compared to one another with a view to the analysis of operation parameters during actual flights of PZL W-3 Sokół and PZL SW-4 Puszczyk, both fitted with on-board flight recorders. Specific emissions generated by the engine at given load states were multiplied by corresponding phase coefficients and then aggregated in order to come up with a single, representative specific emission value.

Table 3.12. Relative time shares of different load ranges for selected helicopters

Helicopter	Test phase range	A	B	C
	$N_e/N_{e\ max}$ [%]	**15**	**70**	**95**
PZL W-3 Sokół	Load time share in overall operation time [%]	1.3	2.7	96
	Test phase share	0.013	0.027	0.960
PZL SW-4 Puszczyk	Load time share in overall operation time [%]	1.0	4.0	95.0
	Test phase share	0.010	0.040	0.950
Mean test phase share		**0.0115**	**0.0335**	**0.9550**
Phase share error relative to the mean value [%]		11.53	24.07	0.52

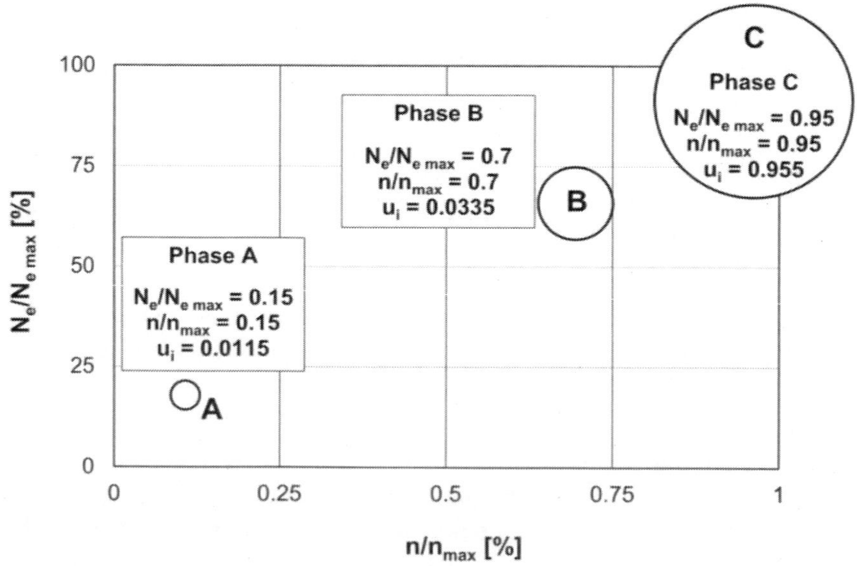

Figure 3.60. Overview of phases and their respective shares in the proposed universal test for helicopters powered with turbine engines.

Table 3.13. Specific emissions of toxic compounds in exhaust gas at different engine loads for selected helicopters

Helicopter	Test phase range	A	B	C	Test
	$N_e/N_{e\ max}$ [%]	15	70	95	–
	u_i [%]	0.0115	0.0335	0.9550	–
PZL W-3 Sokół	e_{CO_2}	3200	1365	1007	1044
	e_{CO}	51	18.6	12.7	13.3
	e_{HC}	22	2.76	1.71	1.98
	e_{NOx}	5.96	2.05	2.09	2.13
PZL SW-4 Puszczyk	e_{CO_2}	4515	1193	1147	1187
	e_{CO}	26.7	19.5	14.2	14.5
	e_{HC}	38.3	2.07	1.25	1.71
	e_{NOx}	0.64	1.27	1.32	1.31
Mi-2 A	e_{CO_2}	3905	1245	1035	1075
	e_{CO}	211	44.1	25.9	28.7
	e_{HC}	31.3	4.71	2.64	3.04
	e_{NOx}	1.55	3.07	2.36	2.37
Mi-2 B	e_{CO_2}	4796	1542	1074	1132
	e_{CO}	213	20.9	11.3	13.9
	e_{HC}	87.1	5.49	1.04	2.18
	e_{NOx}	3.178	1.57	1.75	1.76
Mi-17	e_{CO_2}	1867	1473	1148	1167
	e_{CO}	51.8	18.5	13.8	14.4
	e_{HC}	28.2	2.33	1.17	1.52
	e_{NOx}	5.94	1.59	1.65	1.70

Specific emissions of individual compounds derived from the test were compared against specific emission values of those compounds generated by the engines of respective helicopters in actual operation conditions. As a result of this comparison, a percentage difference between actual specific emissions and test-derived specific emissions was identified. The resulting percentage differences in specific emissions of each compound (actual operation vs. emission test) are additionally presented in Figures 3.61–3.64.

A comparison of specific emission of carbon dioxide generated in actual operation against specific emission determined in accordance with the helicopter emission test indicates that the differences between these two values do not exceed 6.1% (Figure 3.61). In all analyzed helicopters (except for PZL SW-4 Puszczyk) test-derived specific emissions of carbon dioxide are lower than actual emissions of this compound. The greatest difference in the compared values (approx. 6%) is relatively insignificant, which allows one to claim that in terms of energy compliance of the analyzed helicopter engines the assumed test phase coefficients are correct.

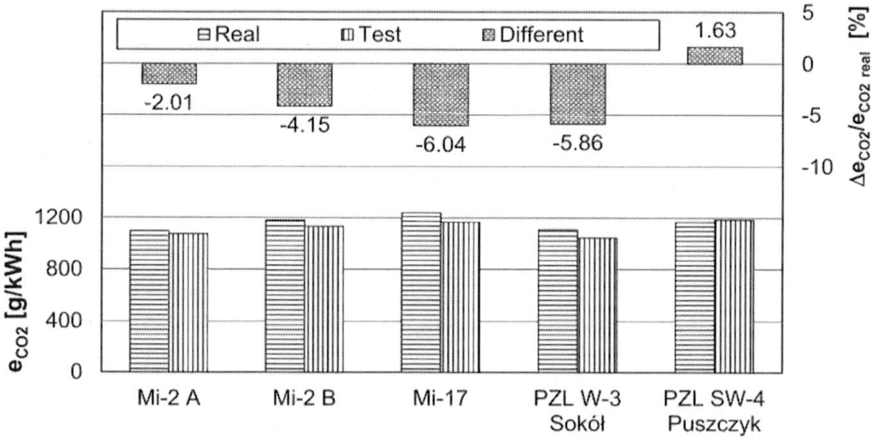

Figure 3.61. Comparison of specific emission of carbon dioxide in actual conditions and in emission test for all study helicopters.

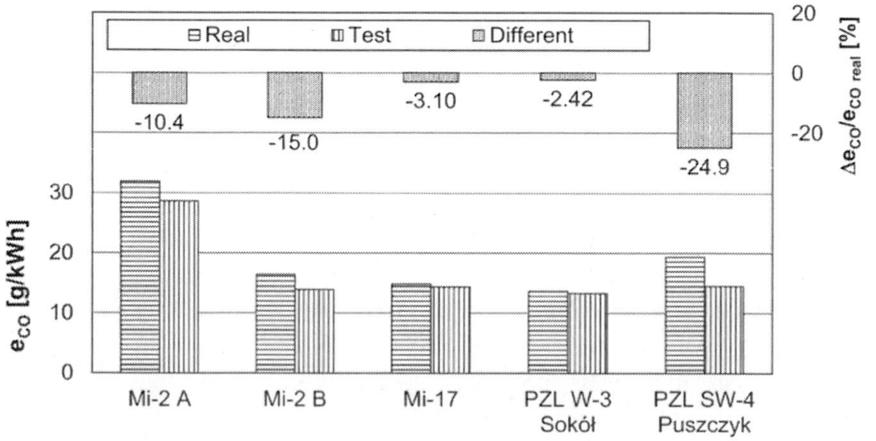

Figure 3.62. Comparison of specific emission of carbon monoxide in actual conditions and in emission test for all study helicopters.

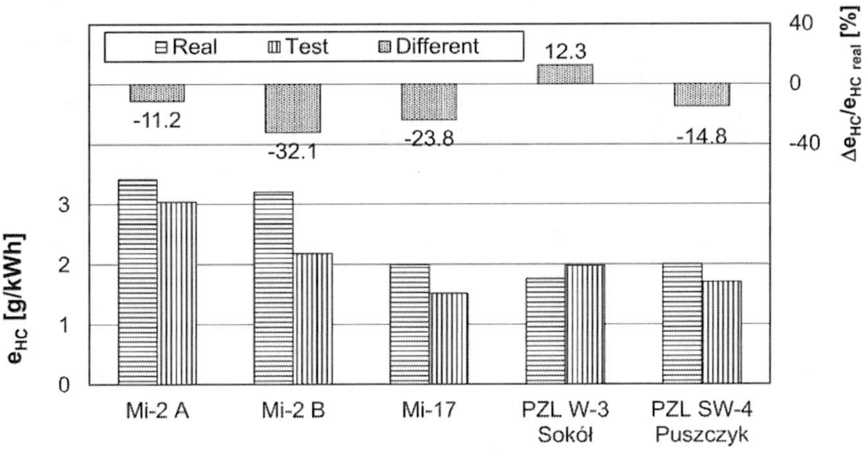

Figure 3.63. Comparison of specific emission of hydrocarbons in actual conditions and in emission test for all study helicopters.

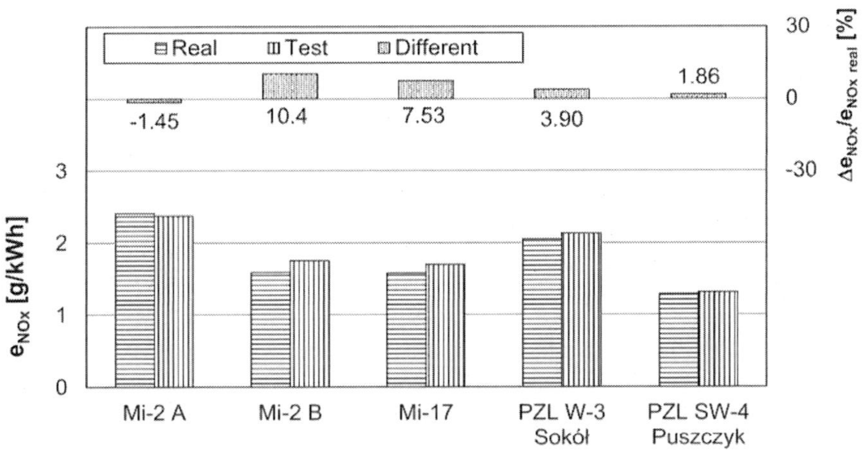

Figure 3.64. Comparison of specific emission of nitrogen oxides in actual conditions and in emission test for all study helicopters.

The other toxic compounds contained in exhaust gas generated by the engines of the tested helicopters were analyzed in a similar way. As regards carbon monoxide, the differences for all helicopters (except for PZL SW-4 Puszczyk) did not exceed 15% (Figure 3.62). For PZL SW-4 Puszczyk the difference between carbon monoxide emissions was approximately 25%.

In the case of hydrocarbons, the differences for all helicopters except for the Mi-2 B did not exceed 25% (Figure 3.63). For the Mi-2 B the difference between carbon monoxide emissions was approximately 32%. In the case of nitrogen oxides, the differences for all helicopters did not exceed 11% (Figure 3.64).

An analysis of the relative differences between specific emissions of individual compounds estimated for actual helicopter operation conditions on the one hand and comparable values determined on the basis of the newly developed emission test on the other shows that most of those differences were relatively small, usually not exceeding 15%. However, three out of all compared pairs of values are characterized by significant differences. These three pairs include: specific carbon monoxide emission from PZL SW-4 Puszczyk (25% difference between actual value and test-derived value); specific hydrocarbons emission from Mi-2 B (32% difference between actual value and test-derived value); and hydrocarbons emission from Mi-17 (24% difference between actual value and test-derived value).

The above analysis of the helicopter emission test compliance shows that the test is designed correctly and is characterized by a high degree of consistency between test-derived specific emissions and specific emissions generated by helicopters in actual operating conditions.

Therefore, it needs to be emphasized that the emission test carried out in accordance with a predefined procedure can be used to assess both emissions and service condition of helicopters. However, in order to obtain information about technical and environmental performance of helicopter propulsion systems it is necessary to determine admissible specific emission limits for all toxic compounds in exhaust gas. The following limits are proposed:

$$e_{CO} = 15 \text{ g/kWh},$$
$$e_{HC} = 2 \text{ g/kWh},$$
$$e_{NOx} = 2 \text{ g/kWh}.$$

If emissions from a given helicopter turbine engine exceed these limits, it could provide an impulse to take measures aimed at a detailed verification of such an engine's technical condition.

Tests and analyses of actual operating conditions of transport helicopters carried out by the Authors confirm major differences in energy and environmental performance of different propulsion systems. The said differences are quite significant, even if adjusted for differences in service lifetimes of individual helicopters.

In order to determine emissions from helicopters, the Authors developed a static test which can be carried out during pre-flight tests and which fairly reflects exhaust gas emissions in actual operation conditions. Static test phase share coefficients used in the test make it possible to adequately assess the engine's environmental condition without the need to carry out time and energy-intensive test bench measurements.

Understanding transport helicopters' environmental impacts involved an analysis of their actual operation conditions and simultaneously required determining the exhaust gas emissions in such conditions. It is possible to do so as long as helicopters' specific operation method is taken into account.

The abridged static emission test for transport helicopters reflects actual helicopter operation conditions and thus makes it possible to determine helicopters' environmental impacts.

More detailed conclusions in several fields can also be drawn from the studies and analyses carried out by the Authors:

- An analysis of histograms showing helicopter loads during actual operation carried out on the basis of data recorded by on-board flight recorders of PZL W-3 Sokół and PZL SW-4 Puszczyk demonstrates that for 95% of flight time helicopter engines work at 80–95% of the maximum load. The remaining 5% of time corresponds to starting up, warming up and cooling down the engine.
- An analysis of measured concentrations of toxic compounds in exhaust gas during pre-flight tests shows that the initial engine start-up phase is characterized by a significant concentration of hydrocarbons, ranging from 300 to 4000 ppm depending on engine type, which is characteristic of turbine engines and closely related to the ignition initiation mode in the combustor. Once the supplied fuel is ignited, carbon monoxide concentration grows rapidly up to 4000 ppm and then is swiftly reduced, down to approximately 1500 ppm for can combustor engines and to approximately 500 ppm for annular combustor engines. Exhaust gas concentration of hydrocarbons also decreases quickly down to approximately 200 ppm. When the engine's thermal state stabilizes, hydrocarbons concentrations stay below 70 ppm. As the engine warms up, the concentration of nitrogen oxides goes up as well. Nitrogen oxides and carbon dioxide concentrations in exhaust gas from a warm engine change depending on engine loads. Therefore, it can be claimed that information on

engine loads can be inferred from records showing the concentration of these two compounds.

- An analysis of specific emissions of toxic compounds in exhaust gas shows that those emissions largely depend on the relative share of a given load value in the total engine operation time, as different toxic compound concentrations correspond to different engine loads. Therefore, although characterized by high concentrations of toxic compounds in exhaust gases, the engine start-up and warm-up phase is not particularly harmful to the environment (low emission). On the other hand, emissions generated at 80–95% of the maximum load are the most significant from an environmental perspective.
- An all-purpose test has been developed, dedicated for assessing emissions from helicopters propelled with turbine engines. The test consists of three phases characterized by different engine loads and different coefficients representing their respective shares in the overall test time. The first phase corresponds to 15% of the maximum engine load and its test time share is 0.0115. The second phase corresponds to 70% of the maximum engine load, with a time share of 0.0335. The last phase corresponds to 95% of the maximum engine load and its share in the overall test time is 0.955.
- A comparison of specific emissions of each compound derived from the aforementioned test against specific emissions generated by the engines of each helicopter in actual operation shows that the relative difference within corresponding pairs of emission values usually does not exceed 15%. However, there are three exceptions: specific emissions of hydrocarbons from Mi-17 differ from each other by 25%; for Mi-2 A the difference between actual hydrocarbons emission value and the test-derived value is approximately 32%; finally, specific carbon monoxide emission values for PZL SW-4 Puszczyk differ from each other by 25%.

A number of general conclusions were formulated after the study:

- To reduce negative environmental impacts of transport helicopters it is necessary to understand actual operation conditions of their engines and to develop representative emission tests enabling energy and environmental performance assessment of helicopter propulsion systems.

- By using helicopter emission tests for assessing environmental performance of helicopter engines one can efficiently reduce time and energy consumption and quickly diagnose the engine's technical condition. By adequately selecting test phase share coefficients one can assess technical condition of new and used helicopter engines.
- The helicopter emission test developed by the Authors should be implemented as an obligatory or recommended test tool for assessing helicopter emissions in actual operation conditions.

REFERENCES

[1] Antas, S. (2006). Determination of performance and parameters for turboprop and turboshaft engine for modification through change of gas temperature before turbine. *Combustion Engines*, *4* (127).

[2] Bickerstaff, C. (2005). Airbus global market forecast – in the context of environment. London: Airbus.

[3] Kotlarz, W. (2004). *Turbine driving systems as pollution sources at military airports*. Dęblin: Air Forces Academy.

[4] Łapucha, R. (2004). Combustion chambers of turbo-jet engines. Warsaw: Scientific Library of the Institute of Aviation.

[5] Merkisz, J., Markowski, J. & Pielecha, J. (2012). Exhaust emissions from the PZL W-3 Sokol helicopter based on the measurements of the concentrations of exhaust components in the exhaust gas during a pre-flight test. Seoul: The 8th IEEE Vehicle Power and Propulsion Conference VPPC 2012, F20120618-0339.

[6] Merkisz, J., Markowski, J., Pielecha, J. & Kozłowski R. (2012). Badania emisyjności śmigłowca Mi-17 w warunkach próby przedstartowej silników. *Logistyka*, *1*.

[7] Merkisz, J., Markowski, J., Pielecha, J. & Kozłowski, R. (2011). Badania emisyjności śmigłowca PZL W-3 Sokół w warunkach próby przedlotowej silników. *Journal of Aeronautica Integra*, *1*.

[8] Merkisz, J., Markowski, J., Pielecha, J. & Kozłowski, R. Emission tests of the TW2-117 helicopter turbine engine during prestart engines test of the MI-17 helicopter. *Logistyka*, *6*.

[9] Merkisz, J., Markowski, J., Pielecha, J., Kozłowski, R. & Mikutel, T. Emission tests of the TWD-10 B/PZL-10S turboprop engine during prestart engines test of the PZL M28B Bryza aircraft. *Logistyka*, *1*.

[10] Merkisz, J., Markowski, J., Pielecha, J., Kozłowski, R. & Mikutel T., Emission tests of the turbopropengine of steady operating conditions. *Scientific Journal of Polish Naval Academy, 1.*

[11] Merkisz, J., Markowski, J., Pielecha, J., Kozłowski, R. & Smuktonowicz S. (2012). Konstrukcja testu badawczego do oceny emisji związków szkodliwych spalin z silników śmigłowca PZL W-3 Sokół. *Journal of Aeronautica Integra, 2.*

[12] Merkisz, J., Markowski, J., Pielecha, J., Mikutel, T. & Kozłowski, R. (2010). Emission tests of the turboprop engine TWD-10 B/PZL-10S of the aircraft PZL M28B Bryza during steady operating conditions. *Zeszyty Naukowe Wyższej Szkoły Oficerskiej Wojsk Lądowych im. gen. T. Kościuszki, 4* (158).

[13] Merkisz, J., Markowski, J., Pielecha, J. & Pielecha I. (2012). Exhaust emissions from the PZL SW-4 Puszczyk helicopter based on the measurements of the concentrations of exhaust components in the exhaust gas during a pre-flight test. *Journal of POLISH CIMAC, Diagnosis, Reliability and Safety, 7* (2).

[14] Merkisz, J., Markowski, J., Pielecha, J. & Smuktonowicz S. (2012). The exhaust emissions measurements from the PZL SW-4 Puszczyk helicopter during a pre-flight test. Zilina: International Conference on Air Transport, INAIR 2012.

[15] Witkowski, R. (2003). *Wprowadzenie do wiedzy o śmigłowcach.* Warszawa: Biblioteka Naukowa Instytutu Lotnictwa.

[16] www.bredow-web.de/.../ a_Turbomotor_TW3-117-d.jpg

Chapter 4

INVESTIGATIONS INTO THE EXHAUST EMISSIONS FROM JET ENGINES

In-flight operation of multi-role aircraft used for military purposes is characterized by highly changeable operating conditions of the propulsion system. As a result, actual operating conditions can hardly be reflected in stationary tests. Therefore, an attempt was taken to analyze engine operation parameters during actual operation. On the basis of information collected in this way an emission test was developed, reflecting operation conditions of those engines. Measurements of emission concentrations during pre-flight tests coupled with the proposed emission test made it possible to determine specific emissions of toxic compounds, based on which environmental performance of multi-role aircraft propulsion systems can be assessed.

4.1. ANALYSIS OF OPERATION CONDITIONS

A multi-role fighter aircraft, being a technical object with a specific function and exhibiting specific operation properties resulting from its design, affects its operating environment both mechanically (by interfering with the medium in which it operates) and chemically (by generating toxic compounds as a result of using consumables and fuels). Unlike civil aircraft, multi-role fighter aircraft are characterized by a high degree of flight parameters' variability. One of the characteristic features of multi-role aircraft is the capability of performing a number of aviation tasks, such as chase, aerial combat at various altitudes, interception of targets, aerial reconnaissance or patrolling. The nature of all these tasks differs from one operation to another, depending on a number of parameters, such as flight altitude, velocity or type of maneuvers. Their complexity while attacking ground targets is illustrated by Figure 4.1.

All flight parameters listed above are reflected in propulsion system performance characteristics, and thus in emissions of toxic compounds contained in exhaust gas. State-of-the-art emission measurement technologies do not allow measuring toxic emissions in actual flight conditions. Collection of exhaust gas samples for analysis and safe installation of measurement equipment are the key challenges in this regard. Therefore, emissions of toxic compounds in exhaust gas are measured in a stationary LTO test. However, tests of this kind do not allow one to determine emission rates of each toxic compound in exhaust gas in actual operation conditions. Therefore, efforts are being made to develop new methods enabling a more accurate assessment of toxic emissions from multi-role aircraft. One of the concepts proposes to assess toxic emissions on the basis of functional connections between operation parameters and emission parameters. Therefore, it is of key importance to analyze operation parameters recorded by on-board flight recorders. Some of the many parameters recorded in this way may be helpful in analyzing operation conditions of the aircraft in general and its propulsion system in particular. For instance, to assess aircraft operation conditions one can use recorded changes in velocity (v), flight altitude (h_{fly}), thrust lever positions (P_p), exhaust gas temperature as well as afterburner on/off information, relative 1st and 2nd stage turbine speeds (n_1 and n_2), relative opening of the exhaust gas exhaust nozzle (WO), and instantaneous fuel consumption (G_e) (Figure 4.2).

Once the operation conditions of an aircraft and its propulsion system have been measured, they can be linked to exhaust gas emission parameters. Subsequently, an analysis aimed at assessing emissions from the aircraft's engine can be carried out. To do so, it is necessary to develop a working methodology.

Due to the properties of available test equipment and to the specific nature of aircraft operation, emissions of toxic compounds in exhaust gas generated by the engine of a multi-role aircraft cannot be carried out in actual flight conditions. Therefore, for a number of selected aircraft actual operation parameters were analyzed, based on which their respective time shares were assessed. The results were used to develop an individualized stationary emission test. Subsequently, emission parameters during typical engine tests were measured. On the basis of toxic compounds' concentrations calculated in this way, emission coefficients were determined and linked to operation parameters in the proposed test. The test is dedicated to multi-role aircraft fitted with low dual flow two-shaft turbine jet engines with an afterburner. Bearing in mind that such aircraft are used for military purposes the Authors chose not to publish their identification data, other than generally known and easily accessible engine information.

The engines of a multi-role aircraft in operation are constantly monitored for key parameters, the values of which are recorded by on-board flight recorders. Out of many data recorded in this way, the following ones were selected for the purpose of analyzing aircraft operation conditions: engine speeds (n_1 and n_2); thrust lever positions (P_p); fuel consumption (G_e); exhaust gas temperature at the 2nd stage turbine; relative opening of the exhaust nozzle; and afterburner on/off information.

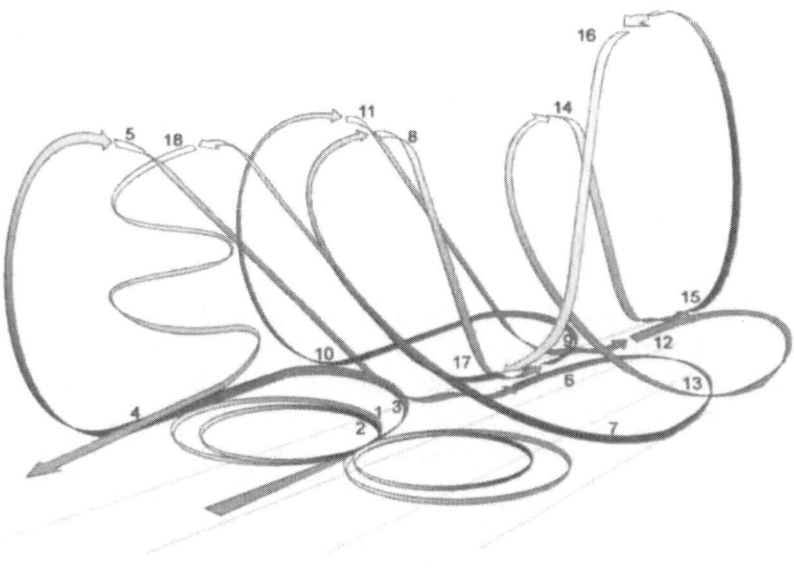

Figure 4.1. Maneuvers of a multi-role aircraft while attacking ground targets [8]: 1 – turn by 45°, 2 – turn by 60°, 3 – half turn left, 4 – roll-off-the-top left, 5 – dive, 6 – half turn right, 7 – roll-off-the-top right, 8 – dive, 9 – half turn left, 10 – roll-off-the-top left, 11 – dive, 12 – half turn right, 13 – roll-off-the-top right, 14 – dive, 15 – Immelmann, 16 – dive, 17 – pull up, 18 – spiral.

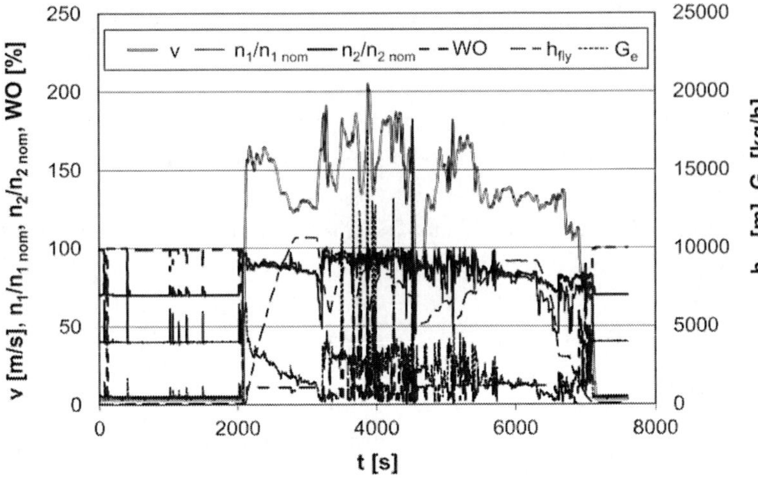

Figure 4.2. Selected operation parameters of a multi-role aircraft and its engine recorded during the flight.

Analyses of operation parameters were carried out for randomly selected flights of several different aircraft. The said parameters were compared against corresponding values from typical engine tests carried out for diagnostic and inspection purposes, such as a pre-flight engine test. On the basis of the results an emission test was developed. The parameters incorporated in the emission test and their respective weights correspond to values recorded during actual operation of multi-role aircraft.

To assess emissions generated by propulsion systems of multi-role aircraft it was necessary to measure concentrations of toxic compounds contained in exhaust gas generated by the engine. Measured concentrations and their corresponding instantaneous engine loads were then compared against operation parameters assigned to individual test points. On that basis, specific emissions of pollutants in test conditions were determined.

Analyzed data were recorded by on-board flight recorders during over ten patrol and training flights. Figure 4.3 shows an example of changes in engine operation parameters recorded during one of such flights. Considering the relative thrust force values (K/K_n) depending on a number of engine operation parameters and present in each operation phase, it is possible to present those values as a function of the relative value of engine speed (n_1). By linking each operation phase to values recorded during the flight, the Authors were able to determine operation areas defined by thrust force as a function of engine speed (n_1).

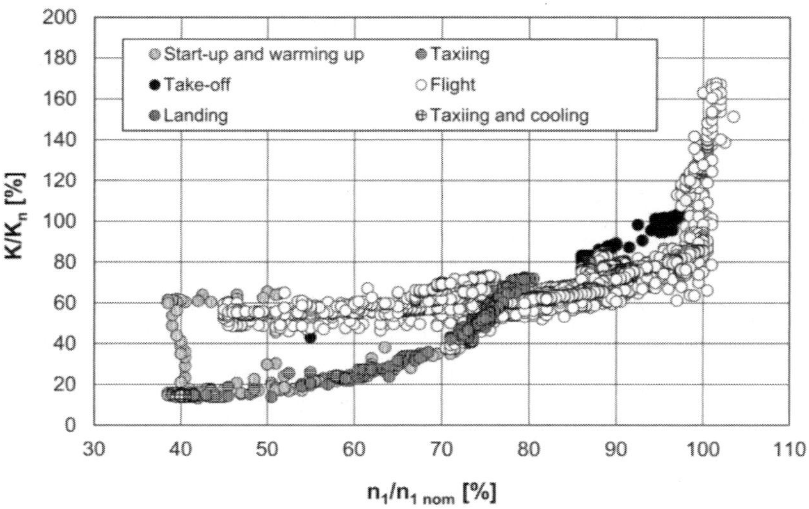

Figure 4.3. Analyzed parameters in the engine operation range, recorded during the flight.

Flight profiles can be described on the basis of the distribution of points corresponding to given phases. For the most part, points corresponding to the engine start-up and warming up phase are located in the area within K/K_{max} = 40% and corresponding to the low-pressure stage speed $n_1/n_{1\ nom}$ = = 70%. A number of points from this particular phase are located outside the said area – they denote the verification of correct operation of exhaust nozzle opening controller. The identified operation area also contains a number of points assigned to the taxiing phase and partially to the landing phase. The latter can be divided into two stages. The first stage corresponds to the approach to land, whereby engine operation parameters stay in the range typical of actual flight (K/K_{max} = 40–80%, $n_1/n_{1\ nom}$ = 70–80%). The other stage is landing, whereby engine parameters are lower (K/K_{max} = 40%, $n_1/n_{1\ nom}$ = 70%). A similar division can be proposed for the take-off phase. In the first stage, i.e., acceleration and take-off proper, one can identify an area corresponding to the range of K/K_{max} = 100% and $n_1/n_{1\ nom}$ = 96%. The other stage corresponds to the ascent after take-off (K/K_{max} = 80–90% and $n_1/n_{1\ nom}$ = 85–90%). The example of relative thrust force and fan speed values during the take-off relates to a plane taking off with the afterburner off and with a small load. If a plane takes off fully loaded, the afterburner is usually off. The afterburner increases the engine's propulsion force by approximately 70%. The extra force comes mostly from the mass of fuel additionally combusted in the afterburner chamber, located downstream of the engine's turbine assembly. The energy generated from fuel delivered to the afterburner chamber increases the energy of exhaust gas ejected from the exhaust nozzle, and thus the overall thrust force. Simultaneously it needs to be emphasized that afterburner status does not affect engine speed parameters (n_1 and n_2). The flow of air through the engine corresponds to values occurring at maximum thrust obtained with the afterburner off. The range of engine operation with the afterburner on covers relative thrust force points in excess of K/K_{max} = 105%. The largest operation area covers points from the flight phase (K/K_{max} = 40–170% and $n_1/n_{1\ nom}$ = 45–105% of the low pressure stage speed). It is evident that this phase contains diverse values of the relative thrust force as a function of engine speed (n_1). The above analysis shows that operation of a low dual flow jet engine fitted in a multi-role aircraft stays in the following ranges: $n_1/n_{1\ nom}$ = 38–105% and K/K_{max} = 40–170%.

The above division of relative thrust force values (K/K_{max}) as a function of relative engine speed ($n_1/n_{1\ nom}$) made it possible to identify the ranges of operation phases. Relative thrust force values below K/K_{max} = 40% and relative engine speed up to $n_1/n_{1\ nom}$ = 70% usually correspond to engine start-up, warming up, taxiing and cooling. Most of the range of K/K_{max} = = 40–100% and $n_1/n_{1\ nom}$ = 45–105% corresponds to take-off and flight, as well as to the approach to land stage of the landing phase. Points in the operation range of K/K_{max} = 100–170% and $n_1/n_{1\ nom}$ = 95–105% correspond to afterburner on situations: these points represent the take-off phase and the maximum engine operation parameters during special maneuvers.

By comparing the K/K_{max} parameters as a function of $n_1/n_{1\ nom}$ for aircraft operation conditions in the air and for pre-flight conditions (Figure 4.4) one notices a remarkable consistency in the distribution of the analyzed parameters. Operation points from the test largely correspond to operation points in actual operation. One can always notice that relative thrust force values during the test correspond to operation conditions during engine start-up, warming up, taxiing and landing, as well as take-off with the afterburner off. The foregoing indicates that the test is optimized in terms of operation time minimization, in line with the assumption that the engine needs to be warmed up and checked for maximum thrust capability with the afterburner off. It also needs to be checked whether the afterburner's operation begins upon its activation. Engine operation values corresponding to K/K_{max} = 40–85% are present during the flight, but are not present during the test.

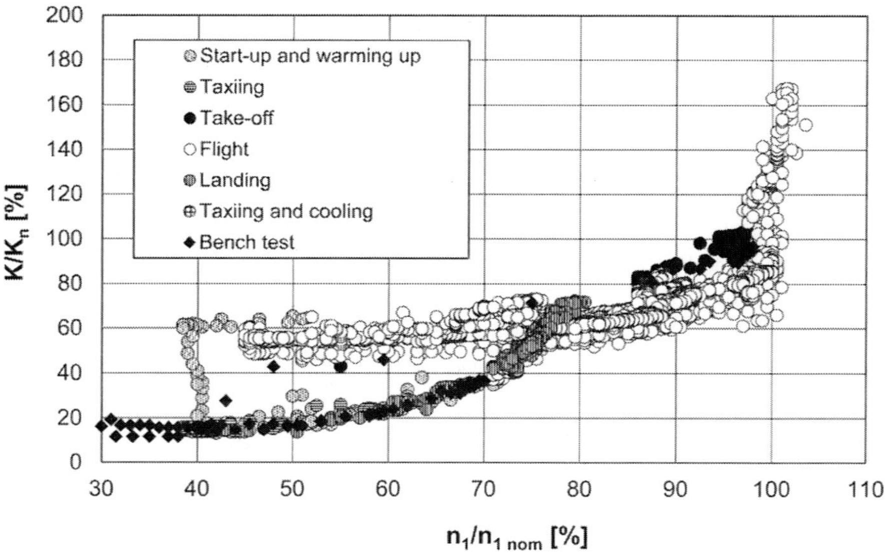

Figure 4.4. Comparison of engine operation parameters recorded during engine test and in flight.

The distribution of analyzed parameters in the engine operation range suggests the possibility to identify characteristic engine operation fields, and to select the afterburner on working range. From the perspective of toxic emissions, the said area is of particular interest, as fuel consumption amounts to approximately 25 400 kg/h, and the combustion process takes place in the afterburner chamber, exhaust nozzle and open space. On the other hand, the area in question is less significant given its small share in the overall engine operation time. Therefore, the afterburner on range was not addressed during pre-flight tests on an engine test bench.

4.2. Stationary Test Design Methodology

The awareness of turbine engine operation criteria and corresponding operation parameter settings for pre-flight test conditions and for actual in-flight conditions makes it possible to develop an all-purpose test for assessing toxic emissions from aviation turbine engines. Once developed and implemented, the test could contribute to reducing the time and effort required for verifying the technical condition of such engines. Therefore, the emission test was designed based on the following assumptions:

- the test should be representative of all multi-role aircraft propelled with turbine jet engines; emission test results from a static test should be comparable to the results of emission measurements in actual operation conditions,
- the number of test phases should not be greater than 3; phases should be selected so as to reflect propulsion system loads typical of actual operation conditions (idling, partial loads, maximum load),
- the test should fit into the scope of pre-flight aircraft tests.

The importance of the third of the above assumptions follows from the fact that pre-flight test duration counts towards aircraft engine service life. Therefore, it would be advisable to avoid reducing the remaining service life by running emission tests separately from pre-flight tests [1, 2–7].

Operation parameter values recorded during selected flights were used as inputs in designing the toxic emission test. All available aircraft propulsion data subjected to statistical analysis were preliminarily divided into three areas: A – $n_1/n_{1\ nom}$ = 35–75% and K/K_{max} = 14–55%; B – $n_1/n_{1\ nom}$ = = 45–100% and K/K_{max} = 50–90%; C – $n_1/n_{1\ nom}$ = 95–105% and K/K_{max} = = 90–100% (Figure 4.5). On the basis of identified ranges of relative thrust forces and relative engine speed (n_1), the values of operation parameters dedicated to measurement points of the emission test were determined (Table 4.2). The values of these parameters correspond to the mean of boundary values of the identified operation ranges of each phase, taking into account time densities of those parameters during actual operation. The assumed coordinates of test points ($n_1/n_{1\ nom}$; K/K_{max}) are as follows: A – (0.70; 0.38), B – (0.85; 0.70), C – (0.95; 0.95). For each determined range of relative engine speeds (n_1), a share in the total engine operation time during the flight was specified. The said shares in the total aircraft flight time are as follows: phase A – 13% of the total flight time; phase B – 57%; and phase C – 30%. In this way it was possible to identify test points at given operation parameters' settings with the corresponding share of a given phase in the total emission of toxic compounds (Table 4.2). A graphical illustration of the test is shown in Figure 4.6.

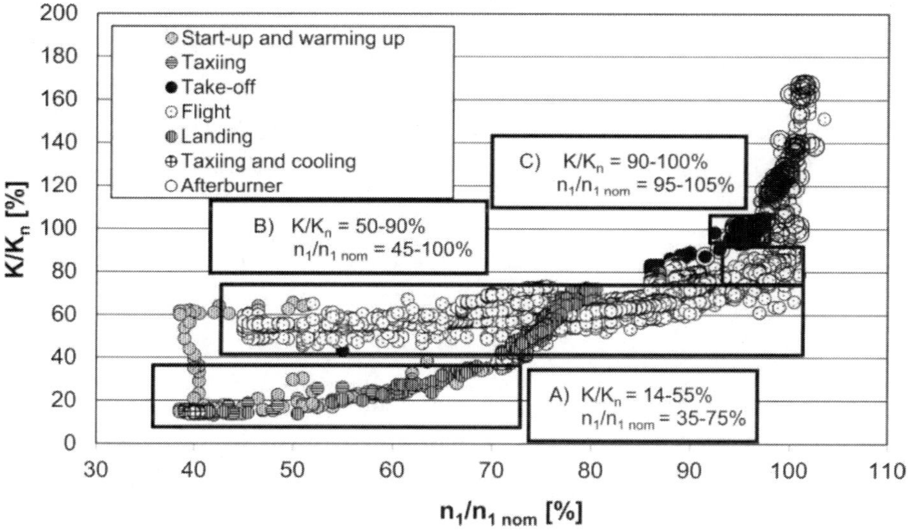

Figure 4.5. Relative engine speed ranges (n_1) broken down into test phases.

Table 4.2. Test phase characteristics and weights

Phase	A	B	C
Phase share	0.13	0.84	0.03
$n_1/n_{1\ nom}$	0.70	0.85	0.95
K/K_{max}	0.38	0.70	0.95

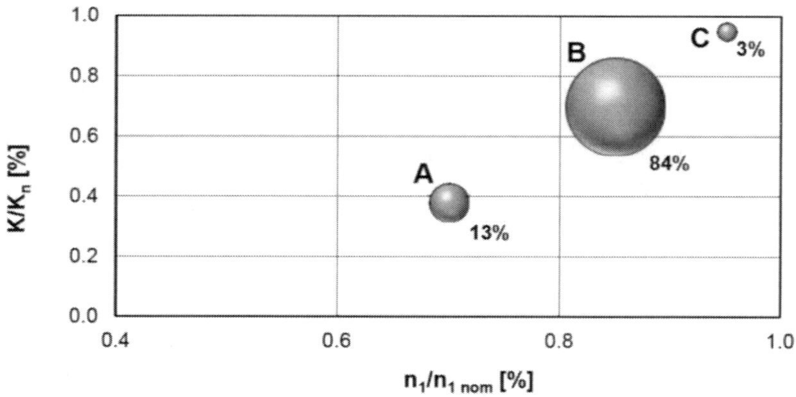

Figure 4.6. Relative shares and characteristic features of phases in the proposed all-purpose static test for turbine engines in multi-role aircraft.

By determining the mean shares of each phase in the total flight time (Table 4.2) and by aggregating the values of hourly emissions in given flight phases multiplied by their corresponding relative shares it is possible to determine the value of emissions of every compound during the entire flight, as per the following equation (4.1):

$$e_j = \frac{\sum E_{ji} \cdot u_i}{\sum K_i \cdot u_i} \text{ [g/kNh]} \tag{4.1}$$

where:

E_{ji} hourly emission of j-th pollutant in test i-th phase i [g/h],
u_i coefficient reflecting the share of i-th phase in the total test time [–],
K_i thrust force in i-th test phase i [kN].

The proposed values of operation parameters in each emission test phase together with their corresponding weights make it possible to determine specific emissions of each pollutant (with reference to thrust force). The proposed emission tests are dedicated to specific engine types. Due to diverse operation ranges of relative engine speeds and the corresponding values of relative thrust force it is not possible to generalize the test and make it suitable for all types of propulsion systems used in multi-role aircraft. The proposed tests should be used separately for single and dual flow jet engines. Assessment of emissions from multi-role aircraft propulsion systems involves measurement of emission rates at specific engine operation points.

4.3. Measurement of Exhaust Gas Pollutants by Means of the Proposed Toxic Emissions Test

The purpose of the study is to measure the emission of pollutants in exhaust gas generated by turbine jet engines used in multi-role aircraft propulsion. Measurements of toxic compounds' concentrations were carried out in a continuous manner for all engine operation parameters corresponding to load values from the engine pre-flight test. Such a test consists in verifying correct operation of all aircraft devices and systems. During a pre-flight test the engine's propulsion performance is assessed. Pre-flight tests are interesting because they cover all engine operation ranges, and measurements of emission performance does not reduce the engine's service time. Therefore, an analysis of procedures connected to aircraft service and operation was carried out in order to develop a method for measuring toxic emissions in exhaust gas during a pre-flight test. An additional advantage of a pre-flight engine test is that its procedure is prescribed by the aircraft manufacturer, and the test itself is obligatory for verifying the engine's technical condition before a flight. The test and the values of all measured engine operation parameters are recorded by an on-board flight recorder fitted in the tested aircraft.

Bearing in mind the above arguments and the results of a comparative analysis of operation parameters' values during the engine test against operation parameters' values in actual (in-flight) operation conditions, measurements of toxic emissions in exhaust gas during such a pre-flight test were carried out. The measured values included concentrations of carbon monoxide, carbon dioxide, hydrocarbons and nitrogen oxides. Measurements of toxic compounds in exhaust gas were recorded simultaneously with engine operation parameters, which made it possible to carry out an in-depth comparative analysis of recorded data and to link engine operation parameters to corresponding concentrations of toxic compounds. Once the values of toxic compounds' concentrations were assigned to engine operation parameters, emission rates accompanying different engine loads were calculated.

Particular attention was paid to the possibility of installing a measurement probe. The probe was installed by means of a metal frame ensuring the probe's permanent position in the axis of the flow of exhaust gas, at a distance of 1 meter from the rim of the engine's exhaust nozzle. The said distance approximately corresponds to the diameter of a single nozzle. All aircraft were tested at test stations dedicated for such purposes, located at the airport. During the test aircraft were immobilized by means of a hook. Ducts used to deliver exhaust gas to the analyzer were fixed to the supporting frame, and the analyzer itself was placed at a safe distance from the exhaust nozzle.

The measurements of toxic emissions in exhaust gas generated by the engine of a multi-role aircraft were carried out during engine tests. Engine operation parameters during the test were recorded by an on-board flight recorder. Selected parameters are presented in Figure 4.7.

The above graph shows the entire duration of a pre-flight test, complete with engine operation range with the afterburner on. The above ranges additionally contain the operation range with the thrust lever in a position exceeding $P_p = 100\%$. The part of the pre-flight test used for the purpose of measuring toxic emissions continued until afterburner activation. The decision to interrupt measurements was taken because of the risk of causing thermal damage to the exhaust gas sampling probe and to the measurement equipment itself. Technically, the probe was removed from the stream of exhaust gas in the following way: after the engine's operation was verified at maximum performance with the afterburner off, engine parameters were shifted down to their minimum values in order to reduce exhaust gas temperature. Subsequently the engine was switched off, and the emission measurement system was dismantled, following which the engine was turned on again in order to complete the remaining parts of the test. Introduction of these changes is visible in operation parameter records registered by the on-board flight recorder during the pre-flight test with emission measurements (Figure 5.8). Propulsion system performance data were recorded as a function of time. Selected toxic compounds in exhaust gas were measured continuously. As a result, it was possible to synchronize individual engine test phases with measured emissions of toxic compounds in exhaust gas.

Bearing in mind the initial part of the pre-flight test record (Figure 5.9) and the presented parameters together with instantaneous values of hourly fuel consumption, they were compared against changes in toxic compounds' concentration in exhaust gas (Figure 5.10).

When looked at against the pre-flight engine test procedure, an analysis of changes in engine parameters made it possible to identify individual engine load states. Individual pre-flight engine test phases (marked with vertical lines) together with a percentage coefficient of the relative thrust force make it possible to read out concentration values in each engine load phase.

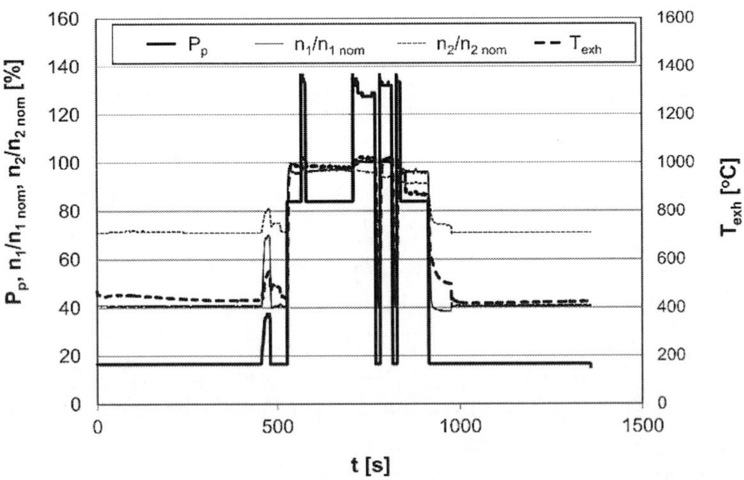

Figure 4.7. Example of a graph illustrating engine operation parameters during a pre-flight test, as recorded by the aircraft's on-board flight recorder.

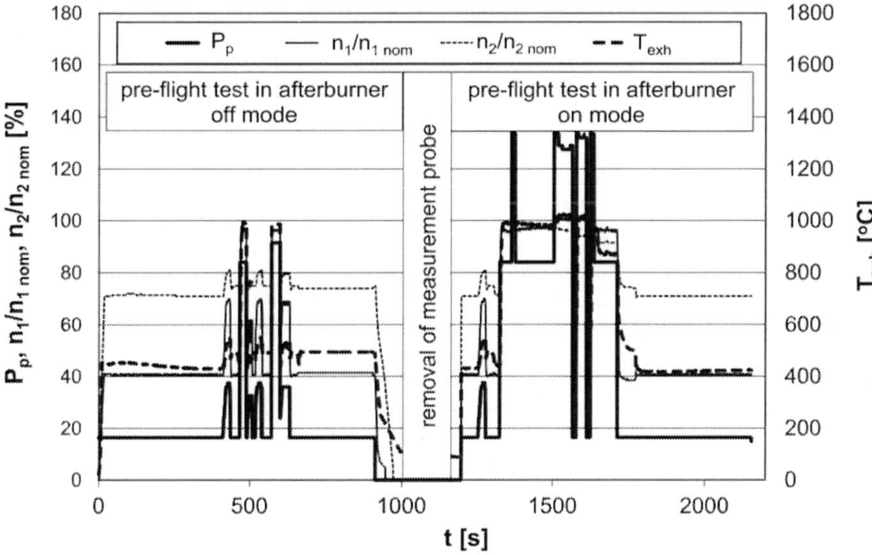

Figure 4.8. Engine operation parameters recorded by the aircraft's on-board flight recorder during pre-flight test with simultaneous measurement of toxic compounds' emissions.

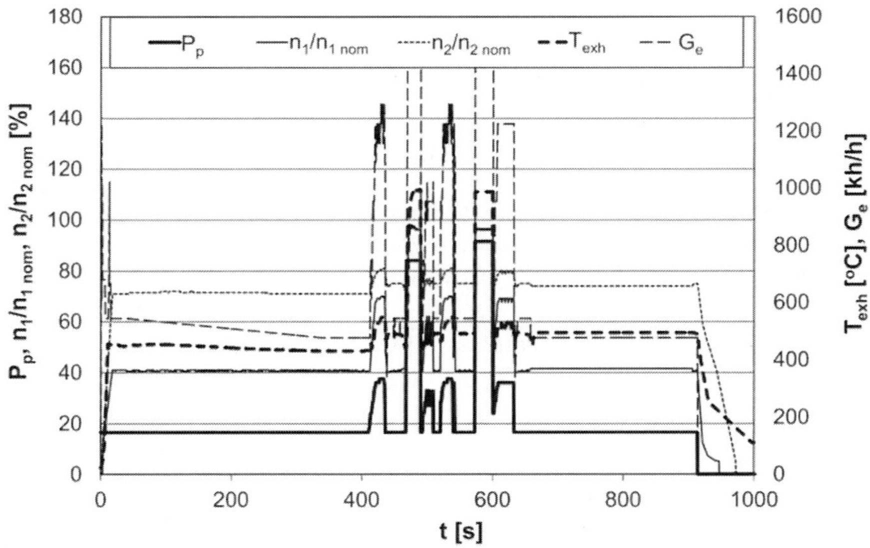

Figure 4.9. Initial part of the pre-flight test with simultaneous measurement of toxic compounds' emissions.

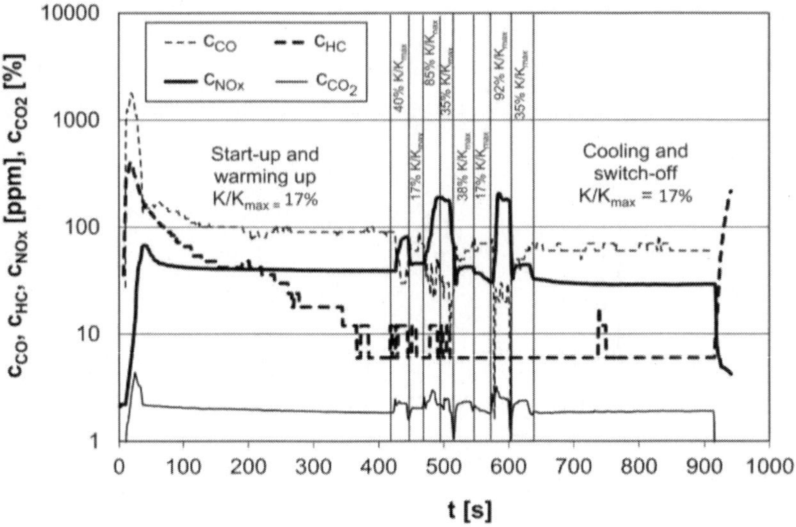

Figure 4.10. Results of toxic compounds' **concentration measurement** as a function of time during a pre-flight engine test.

The beginning of the start-up phase of a turbine engine is highly characteristic. In this part of the test, attention is drawn by a high concentration of hydrocarbons, which directly depends on the mass of fuel fed into the combustor. Subsequently, when ignition in the combustor is initiated, concentrations of carbon monoxide, carbon dioxide and hydrocarbons grow rapidly. Greater concentrations of these compounds are a consequence of the combustion process, the efficiency of which is low in the beginning. As the combustor is warmed up and the target thermal state of other engine elements is achieved, concentrations of these compounds are gradually reduced.

Warming up of the combustor enhances thermodynamic parameters of the combustion process, and thus increases the concentration of nitrogen oxides in exhaust gas. As engine temperature grows, the concentration of nitrogen oxides stabilizes at approx. 45 ppm with engine load corresponding to approximately $K/K_{max} = 17\%$ and resultant engine speed of approx. $n_1/n_{1\ nom} = 70\%$. When engine load grows up to $K/K_{max} = 40\%$ and engine speed grows up to $n_1/n_{1\ nom} = 80\%$, the concentration of nitrogen oxides in exhaust gas increases to approx. 80 ppm. Increase in thrust force up to $K/K_{max} = 85\%$ with engine speed going up to $n_1/n_{1\ nom} = 98\%$ results in further increase of nitrogen oxides concentration in exhaust gas, up to approx. 190 ppm. Increase in engine load up to K/K_{max} = 92%, accompanied by engine speed becoming $n_1/n_{1\ nom}$ = 98% increases the concentration of nitrogen oxides in exhaust gas up to approx. 200 ppm. Changes of similar nature are observed in the case of carbon dioxide. Maximum values of carbon dioxide concentrations at each engine operation point are as follows: for K/K_{max} = 17% and $n_1/n_{1\ nom}$ = 70%, CO_2 = 2%; for K/K_{max} = 40% and $n_1/n_{1\ nom}$ = 80% – CO_2 = 2.3%; and for K/K_{max} = 92% and $n_1/n_{1\ nom}$ = 98% – CO_2 = 2.5%. The concentration of carbon monoxide in exhaust gas reaches its maximum value (1700 ppm) in the initial phase of engine start-up, after initiation of ignition in the combustor. Subsequently it decreases gradually and remains in the range from 50 to 200 ppm. Changes in hydrocarbon concentrations are similar. Maximum HC concentration (400 ppm) is observed during engine start-up, and then as the combustor temperature goes up and the engine reaches its target thermal state, hydrocarbons concentrations go down to approx. 25 ppm. Relatively low concentrations of all compounds are related to a high excess air coefficient in the combustor, which – in the case of turbine engines – results in a certain degree of exhaust gas dilution.

Pairing of load histograms with corresponding emission rates of toxic compounds in exhaust gas made it possible to assess aircraft emissivity in the proposed stationary test, and thus to evaluate emissions in actual operation conditions.

4.4. Assessment of Toxic Emissions from Multi-Role Aircraft

To determine toxic emissions in exhaust gas generated by a turbine jet engine by means of the proposed test it is necessary to specify the value of exhaust gas stream flow in given operation conditions. The mass of toxic gas flow and information on the concentration of toxic compounds in that flow allow one to determine actual emissions of these compounds. The mass of the exhaust gas ejected from the exhaust nozzle was calculated on the basis of the excess air coefficient measured in the gas flow axis. By measuring the value of excess air coefficient in exhaust gas and the value of fuel consumption at a given operation point (Table 4.4) one can calculate the emission rate of exhaust gas ejected from the engine. Concentrations of each of the compounds in exhaust gas at predefined operation points are shown in Figure 4.11.

Table 4.4. Values of excess air coefficient in exhaust gas, fuel consumption, exhaust gas flow and toxic compounds' concentration for given test phase parameters

Parameter	Phase		
	A	B	C
Thrust force [kN]	30.1	55.4	75.1
Phase share coefficient	0.13	0.84	0.03
Excess air coefficient in exhaust gas [–]	8	7	5
Fuel consumption [kg/h]	550	1100	5897
Exhaust gas flow [kg/h]	80 630	141 240	542 524

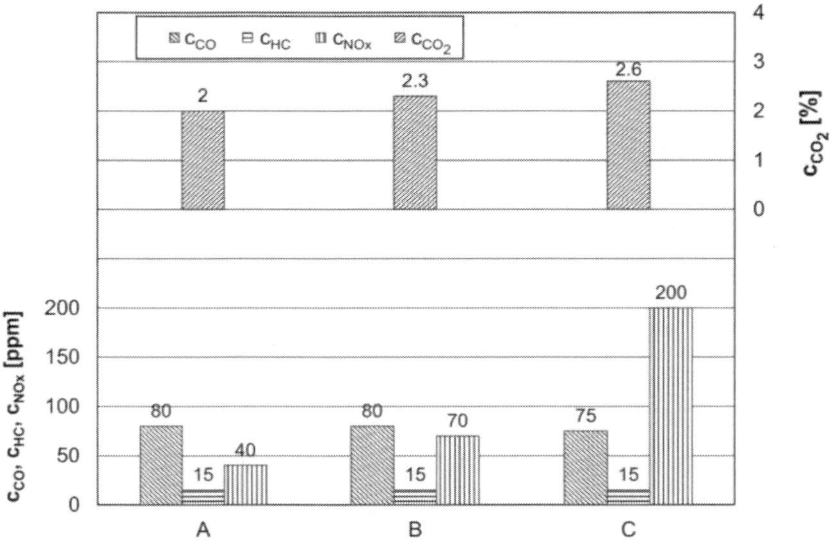

Figure 4.11. Concentrations of toxic compounds' **emission** in exhaust gas in given test phases.

By identifying the flow of exhaust gas and the concentration of toxic compounds, emission rates of toxic compounds in exhaust gas were calculated for different engine operation points (Figure 4.12).

By applying the equation presented above (4.1), specific emissions of toxic compounds in exhaust gas were calculated for the purpose of the proposed emission test with respect to aircraft operation conditions (Figure 4.13).

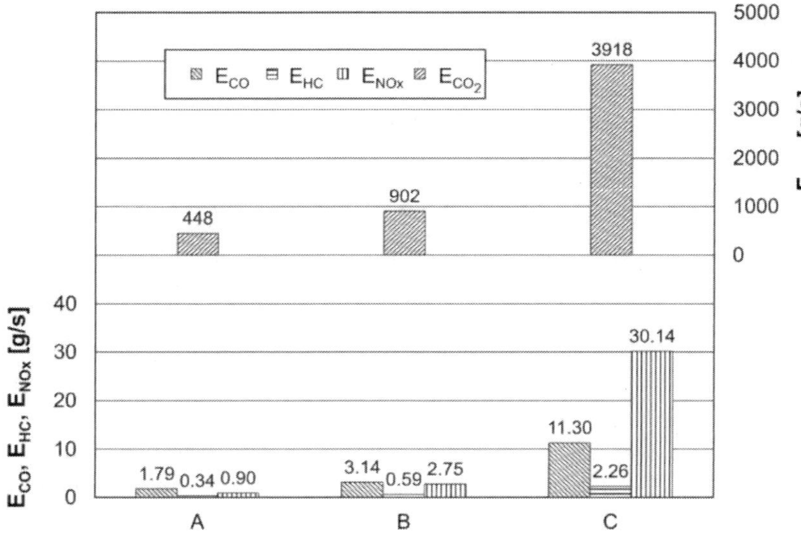

Figure 4.12. Emission rates of toxic compounds in each test phase.

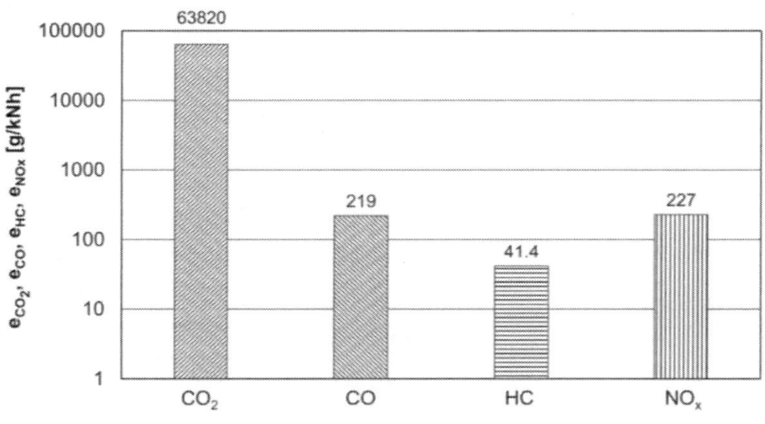

Figure 4.13. Specific emission of toxic compounds in exhaust gas generated by aircraft engine calculated on the basis of the proposed test.

The assessment of emissions of toxic compounds in exhaust gas generated by the engine of a multi-role aircraft constitutes an attempt at developing testing methodology aimed at using toxic emission measurements for the purpose of diagnosing the engine's technical condition. Information obtained in this way can be helpful in verifying and developing test procedures. The above study should be treated as preliminary, and its results should be correlated with results obtained for several aircraft of the same type. Eventually, such studies may contribute to developing generally applicable testing procedures determining the emissivity of aircraft and their impact on the environment.

REFERENCES

[1] Kotlarz, W., Piaseczny, L., Rypulak, A. & Zadrąg, R. (2006). Tests of exhaust gas toxicity of jet turbine engine for take-off and landing phases of flight. *Combustion Engines*, *4* (127).

[2] Merkisz, J., Markowski, J. & Pielecha, J. (2013). Emission tests of the F100-PW-229 turbine jet engine during pre-flight verification of the aircraft. In J. W. S. Longhurst, & C. A. Brebbia (Eds.), *WIT Transactions on Ecology and the Environment*, (*174*, 219-230). Southampton: WIT Press.

[3] Merkisz, J., Markowski, J. & Pielecha, J. (2013). Emission tests of the F100-PW-229 turbine jet engine during prestart trial of the aircraft. Guilin, China: The 26th International Conference on Efficiency, Cost, Optimization, Simulation and Environmental Impact of Energy Systems.

[4] Merkisz, J., Markowski, J., Pielecha, J. & Smuktonowicz, S. (2011). Ocena emisji związków szkodliwych spalin silnika odrzutowego F100-PW-229 podczas próby na stanowisku hamownianym. *Logistyka*, *6*.

[5] Merkisz, J., Markowski, J., Pielecha, J. & Smuktonowicz, S. (2011). The measurement of the exhaust emissions from the F100-PW-229 turbine jet engine during a preflight engine test of the fighter plane. *Journal of KONBIN*, *1*.

[6] Merkisz, J., Markowski, J., Pielecha, J. & Usarek W. (2012). Konstrukcja testu stacjonarnego do oceny emisyjności samolotu. In T. Łodygowski, & M. Ciałkowski (Eds.), *Rozwój techniki, technologii i transportu w lotnictwie*. Poznań: Wydawnictwo Politechniki Poznańskiej.

[7] Merkisz, J., Markowski, J., Pielecha, J. & Usarek W. (2012). Ocena rozkładu wymiarowego cząstek stałych w spalinach mikro silnika turbinowego. In T. Łodygowski, & M. Ciałkowski (Eds.), *Rozwój techniki, technologii i transportu w lotnictwie*. Poznań: Wydawnictwo Politechniki Poznańskiej.

[8] Wygonik, P. (2006). Selection criteria of turbine engine parameters for multi-purpose aircraft. *Combustion Engines*, *4* (127).

Chapter 5

CONCLUSION

Increasingly strict regulations on emission levels in exhaust gas are one of the key drivers stimulating the development of combustion engine designs. Determined thus far by the need to meet emission limits applicable to two key exhaust gas ingredients (nitrogen oxides and particulate matter), the said development involves constant searching for new ways of reducing emissions of those compounds. One of the effects of such efforts is the introduction of emission limits for all means of transportation propelled with internal combustion engines. In the case of aircraft, the above measures are limited first of all to high power or high thrust force turbine engines. As a result, current limits do not apply to piston and turboshaft engines used in airscrew propulsion systems dedicated to small airplanes and helicopters. Consequently, the development of the designs of such aviation engines has been stalled, which means that their operation causes major environmental hazards. Particularly important here is the emission of carbon monoxide and nitrogen oxides, i.e., compounds that constitute a direct threat to human life and to ozone layers in higher parts of the atmosphere (nitrogen oxides). In the absence of precise data on toxic emissions from aviation piston engines, a detailed analysis of those emissions was carried out, accompanied by an analysis of toxic emissions from turbine engines used in helicopters and in jet propulsion systems of multi-role aircraft.

The objective behind the said efforts was to develop a tool for determining emission levels from selected aviation propulsion systems in actual operation conditions. To do so, it was necessary to analyze operation parameters of selected systems. Stationary emission tests developed in this way measure engine operation parameters at predefined measurement points.

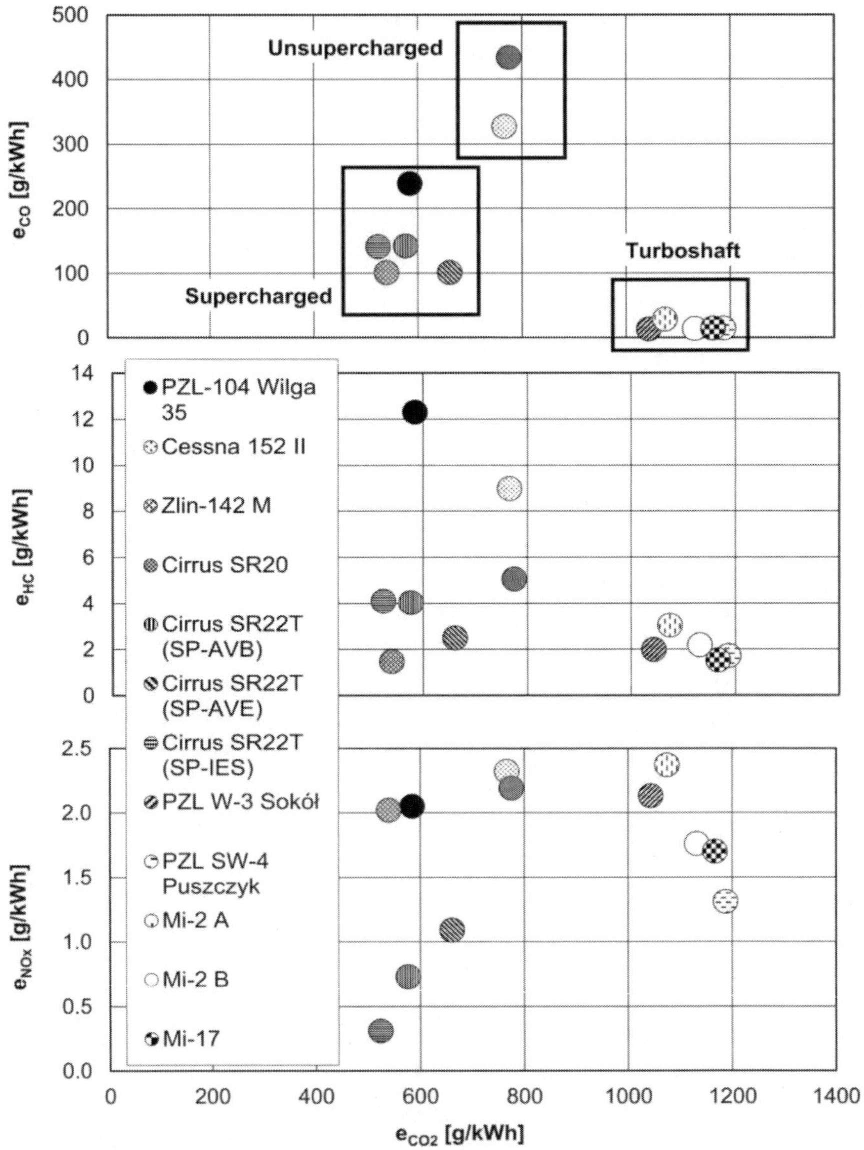

Figure 5.1. Relationship between specific emission of pollutants from aviation piston engines and specific emissions of carbon dioxide.

Paired with specific power or thrust force, the results of toxic emissions tests can be used for comparing engines' technical condition or for determining their environmental performance. Specific emissions of a given compound shown as a function of specific emission of carbon dioxide (Figure 5.1) can be used to generalize aircraft emission ranges determined by characteristic features of their propulsion systems.

The above specific emission ranges for piston engine airplanes depend on the properties of their engines. On the one hand, specific emission of carbon dioxide (linked to fuel consumption) may indicate the propulsion system's efficiency. On the basis of the charts presented above one can conclude that piston engine propulsion systems are more efficient than turbine engine systems used in helicopters. On the other hand, specific emission of toxic compounds depends on the excess air coefficient, and – for piston engines – it is closely related to fuel and air mixture composition and to engine supercharging. In supercharged piston engines the above features contribute to lower specific emissions of carbon monoxide and hydrocarbons, as compared to unsupercharged engines. When compared to turbine engines, specific emission of carbon monoxide from the tested piston engines is 10 or 60 times greater (supercharged or unsupercharged engines, respectively).

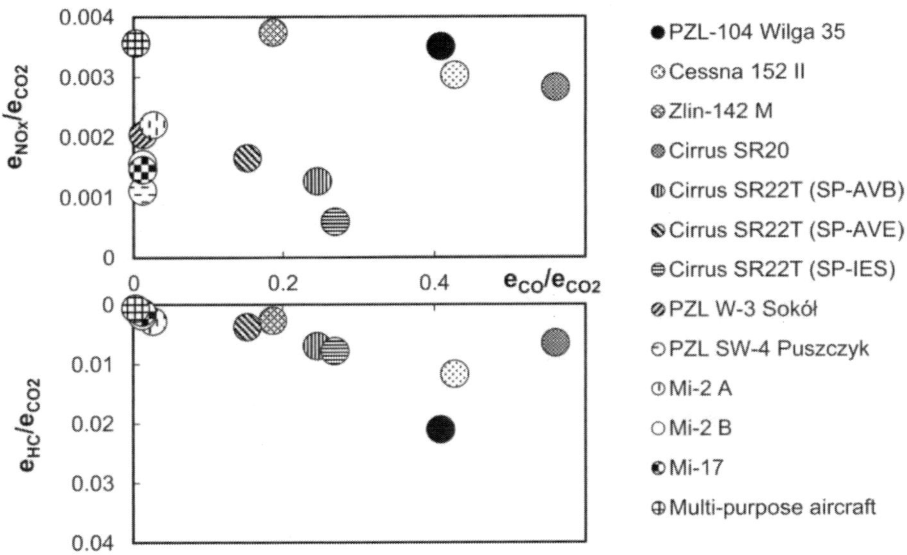

Figure 5.2. Dimensionless emission coefficients for all types of aviation engines.

Since emissions were assessed with the use of different aircraft, the previous comparison does not include emission characteristics of multi-role airplanes. To do so, the value of specific emission of each compound was divided by the value of specific emission of carbon dioxide (expressed in relevant units). In this way, dimensionless emission coefficients for all propulsion types (including jet propulsion) were arrived at (Figure 5.2).

Emission coefficient values shown in the charts indicate a greater threat of toxic emissions from piston engines. Unsupercharged piston engines show particularly negative properties, characterized by maximum emissions of carbon monoxide and hydrocarbons, and by elevated emissions of nitrogen oxides. Aircraft propulsion systems based on turbine engines (including the jet engine) are characterized by lower emissions of these compounds.

Issues presented above highlight the problem of increased emission of toxic compounds from aviation piston engines. Specific emissions of those compounds may constitute a starting point for introducing prescribed limits of such emissions. They can also contribute to greater interest in mathematical modeling for assessing emissions from aircraft in actual operation conditions.

AUTHORS' CONTACT INFORMATION

Dr. Jerzy Merkisz
Head of Institute of Combustion Engines
and Transport at Poznan
University of Technology
Poland
E-mail: Jerzy.Merkisz@put.poznan.pl

Dr. Jaroslaw Markowski
Assistant Professor
Institute of Combustion Engines
and Transport at Poznan
University of Technology
Faculty of Machines and Transport
Poland

Dr. Jacek Pielecha
Assistant Professor
Institute of Combustion Engines
and Transport at Poznan
University of Technology
Faculty of Machines and Transport
Poland

INDEX

G

H

I

J

K

L

M

N

O

P

R

S